Cerebrospinal Fluid

Edited by Pınar Kuru Bektaşoğlu and Bora Gürer

Published in London, United Kingdom

IntechOpen

Supporting open minds since 2005

Cerebrospinal Fluid
http://dx.doi.org/10.5772/intechopen.95715
Edited by Pınar Kuru Bektaşoğlu and Bora Gürer

Contributors
Hüseyin Yakar, Hayriye Soytürk, Murat Yılmaz, Cansu Önal, Eylem Suveren, Ümit Kılıç, Rohit Gulati, Sadhika Sood, Daniel Jaramillo-Velásquez, Fernando Hakim, Martina González, Diego F. Gómez, Juan F. Ramón, Mateo Serrano-Pinzón, Sarang Gotecha, Ashish Chugh, Prashant Punia, Neelesh Kanaskar, Roger Bayston, Bora Gürer, Pinar Kuru Bektaşoğlu

First published in London, United Kingdom, 2022 by IntechOpen
IntechOpen is the global imprint of INTECHOPEN LIMITED, registered in England and Wales, registration number: 11086078, 5 Princes Gate Court, London, SW7 2QJ, United Kingdom
Printed in Croatia

British Library Cataloguing-in-Publication Data
A catalogue record for this book is available from the British Library

Additional hard and PDF copies can be obtained from orders@intechopen.com

Cerebrospinal Fluid
Edited by Pınar Kuru Bektaşoğlu and Bora Gürer
p. cm.
Print ISBN 978-1-83969-695-4
Online ISBN 978-1-83969-696-1
eBook (PDF) ISBN 978-1-83969-697-8

For EU product safety concerns:
IN TECH d.o.o., Prolaz Marije Krucifikse Kozulić 3, 51000 Rijeka, Croatia, info@intechopen.com or visit our website at intechopen.com.

Meet the editors

Dr. Kuru Bektaşoğlu graduated from the School of Medicine, Marmara University, Istanbul, Turkey, in 2014. She completed a neurosurgical residency in 2021 and obtained a Ph.D. in Physiology from Marmara University Institute of Health Sciences. For six months, she worked as a research fellow at the Department of Neurological Surgery, Yeditepe University, Istanbul, Turkey, where she studied white matter fiber dissection. Dr. Bektaşoğlu has published forty papers in international and national journals and presented them at numerous international and national meetings. She has written one book and eight book chapters. She has a keen interest in neurovascular and neurooncological surgeries. She would like to pursue her research career in translational neuroscience with a special emphasis on vascular and tumor biology.

Dr. Gürer graduated from the Faculty of Medicine, Dokuz Eylul University, İzmir, Turkey, in 2007. He completed his neurosurgical residency in 2013. For six months, he worked as a research fellow in the Department of Neurological Surgery, University of Wisconsin, USA, where he studied cranial base and neurovascular surgery. In 2016, he became the youngest associate professor of neurosurgery in his country. Until 2021 he worked as an educational supervisor at the University of Health Sciences, Istanbul, Turkey. In May 2021, he was appointed Professor of Neurosurgery, Department of Neurosurgery, Istinye University, Istanbul, Turkey. Dr. Gürer has published more than eighty papers in international and national journals. He has also written three books and seven book chapters. He has a keen interest in complex neurovascular, neuro-oncological, and skull base surgeries. He is currently continuing his clinical practice at VM Medical Park Mersin Hospital, Turkey.

Contents

Preface

Cerebrospinal fluid is produced and absorbed at constant rates, maintaining ionic homeostasis. Any disruption in this well-regulated system, such as overproduction, decreased absorption, or obstruction, could lead to hydrocephalus.

This book includes essential knowledge about the anatomy of brain ventricles, physiology, and history of cerebrospinal fluid, main pathologies related to cerebrospinal fluid and ventricles such as infection and intraventricular hemorrhage as well as related treatment strategies. Special consideration is given to normal pressure hydrocephalus. As our understanding of cerebrospinal fluid homeostasis evolves, so too does our ability to develop future novel treatments for managing cerebrospinal fluid-related pathologies.

Pinar Kuru Bektaşoğlu, MD, Ph.D.
Neurosurgeon at University of Health Sciences,
Istanbul Fatih Sultan Mehmet Education and Research Hospital,
Department of Neurosurgery,
İstanbul, Turkey

Bora Gürer, MD
Professor of Neurosurgery,
Istinye University School of Medicine,
Department of Neurosurgery,
İstanbul, Turkey

Section 1

Anatomy, Physiology and History of Cerebrospinal Fluid

Chapter 1

History, Anatomy, Histology, and Embryology of the Ventricles and Physiology of the Cerebrospinal Fluid

Pinar Kuru Bektaşoğlu and Bora Gürer

Abstract

Cerebrospinal fluid is an essential, clear, and colorless liquid for the homeostasis of the brain and neuronal functioning. It circulates in the brain ventricles, the cranial and spinal subarachnoid spaces. The mean cerebrospinal fluid volume is 150 ml, with 125 ml in subarachnoid spaces and 25 ml in the ventricles. Cerebrospinal fluid is mainly secreted by the choroid plexuses. Cerebrospinal fluid secretion in adults ranges between 400 and 600 ml per day and it is renewed about four or five times a day. Cerebrospinal fluid is mainly reabsorbed from arachnoid granulations. Any disruption in this well-regulated system from overproduction to decreased absorption or obstruction could lead to hydrocephalus.

Keywords: arachnoid villi, cerebrospinal fluid pressure, choroid plexus, hydrocephalus, ventricular system

1. Introduction

Cerebrospinal fluid (CSF) is located in the brain ventricles, the cranial and spinal subarachnoid spaces [1, 2]. It acts as a cushion and plays a significant role in brain development, in the regulation of the interstitial fluid of the brain parenchyma and healthy neuronal functioning [1]. CSF is mainly produced from the choroid plexuses and mainly absorbed through arachnoid villi. The mean CSF volume is 150 ml, with 125 ml in subarachnoid spaces and 25 ml in the ventricles. In this chapter, historical understanding, anatomy, histology, embryology of ventricles, and physiology of CSF will be discussed.

2. Historical understanding of the ventricular system and cerebrospinal fluid

In ancient times, ventricles were thought to be the site of emotions, mind, judgment, and memory. Hippocrates (460–375 BC), described congenital hydrocephalus as 'water' around the brain [3]. Most likely, Aristotle (384–322 BC) was the first one who noticed the presence of brain cavities, especially the lateral ventricles [4]. A Greek physician, Herophilos of Chalcedon (335–280 BC) was the true discoverer of the first human cadavers dissections [5]. He also described the choroid plexuses [6]. Erasistratus of Ceos (304–250 BC), a scholar of Herophilus, suggested the ventricular theory [7]. Rufus of Ephesus (110–180), master of Galen of Pergamum, elaborated the lateral, third, and fourth ventricles and the mesencephalic aqueduct [7]. Galen (130–200) also described the ventricular system detailly and mentioned pneuma, a breath that arises from the cosmos which circulates through the brain cavities, and serves as a mediator between body and soul [7, 8]. He also specified that obstruction of the ventricular system causes seizures. In Anathomia (1316), Mondino de Luzzi (1270–1326) preserved the tricameral theory for cerebral ventricles, which is mostly influenced by Galenic tradition [9]. Leonardo da Vinci (1452–1519) made the first ventriculography on the ox brain and extracted a three-dimensional template that showed the shape of the ventricular labyrinth [2, 10]. In 1859, a German anatomist and surgeon Benedict Stilling (1810–1879) describe the terminal ventricle as a cystic cavity lined by ependymal cells located in the conus medullaris for the first time [7]. Then, in 1875, Krause called it the fifth ventricle, named after him as Krause's ventricle [7].

Cerebrospinal fluid is discovered by E. Swedenborg (1688–1772) [11]. In 1747, a Swiss physician A. von Haller (1708–1777), presented that in the brain the 'water' is secreted into the ventricles and absorbed in the veins. Hydrocephalus has resulted from excess secretion, which descends to the skull base and into the "spinal marrow" [12]. Domenico Felice Antonio Cotugno [13] was the one who first defined the connection between cerebral ventricles and subarachnoid space. A French physiologist, F.J. Magendie (1783–1855) also confirmed this finding [13]. An opening in the roof of the fourth ventricle, foramen Magendie, was discovered by Magendie, however, erroneously mentioned that CSF was secreted by the pia mater [14]. T. Willis (1621–1675), an English physician, described "a liquid" in the aqueduct of Sylvius that connects the ventricles, and continued that the consistency of the "liquid" is altered in "epidemic fever," i.e., meningitis [15]. In 1891, W.E. Wynter (1860–1945) by tapping the spinal subarachnoidal space treated tuberculous meningitis [16]. H. Quincke (1842–1922), a German internist and surgeon popularized lumbar puncture and advocated its use for diagnostic and therapeutic reasons [17]. In 1912, a neurologist W. Mestrezat (1883–1928) described the chemical composition of the CSF accurately [18], and in 1914, a pioneer neurosurgeon H.W. Cushing (1869–1939), made it clear that the CSF is secreted by the choroid plexus [19].

3. Anatomy of the ventricular system

In the brain, there are 4 ventricles: 2 lateral ventricles, the third ventricle in the diencephalon, and the fourth ventricle in the hindbrain (**Figure 1**) [2]. It is continuous with the central canal of the spinal cord caudally.

3.1 The lateral ventricles

The lateral ventricle is a C-shaped cavity with a capacity of 7–10 ml [2]. It encompasses the thalamus and diencephalon and is divided into five segments. The lateral ventricle has body (central portion), atrium (trigone), and 3 horns (cornua); anterior (frontal), posterior (occipital), and inferior (temporal) horns [21, 22]. The corpus callosum forms the roof of the lateral ventricle, and the posterior portion of the septum pellucidum lies medially. Septum pellucidum is a thin vertical sheet of nervous tissue covered with ependyma on both sides of the ventricles. The caudate nucleus, the lateral dorsal surface of the thalamus, the anterior part of the body of the fornix, the choroid plexus, and stria terminalis form the floor of the lateral ventricle.

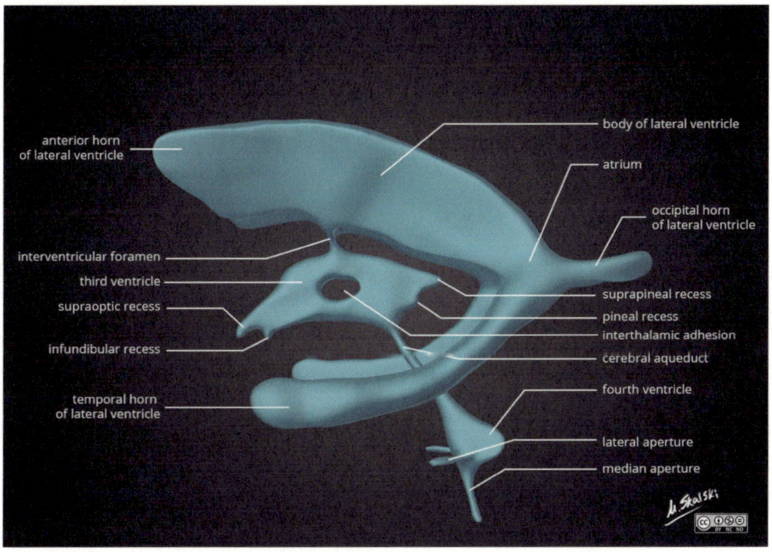

Figure 1.
Illustration of ventricular system anatomy [20].

3.1.1 The body of lateral ventricle (central part)

The body of the lateral ventricle lies within the parietal lobe [2]. The anterior limit is the interventricular foramen and the posterior limit is the splenium of the corpus callosum. The inferior surface of the body of the corpus callosum forms the roof. Mostly septum pellucidum forms the medial wall and in the lower part of the medial wall, there is the body of the fornix. From medial to lateral; choroid fissure, choroid plexus that invaginate into the lateral ventricle through a slit space between the fornix and upper surface of the thalamus, the lateral part of the superior surface of the thalamus, thalamostriate vein, stria terminalis, and the body of the caudate nucleus forms the concave floor.

3.1.2 The horns of the lateral ventricle

3.1.2.1 The anterior (frontal) horn

The frontal horn is located anterior to the interventricular foramen and moves anteriorly and slightly lateral and downward to lie in the frontal lobe [2]. It has an anterior and medial wall, a roof, and a floor. The posterior surface of the genu of the corpus callosum and the rostrum forms the anterior wall. The medial wall is formed by the septum pellucidum. The roof is formed by the inferior surface or anterior part of the body of the corpus callosum. By a majority, the floor is formed by the head of the caudate nucleus, while the upper surface of the rostrum of the corpus callosum forms a small portion on the medial side.

3.1.2.2 The posterior (occipital) horn

Posterior horn turn inversely and medially to lie in the occipital lobe [2, 21, 22]. It is mostly asymmetrical. The tapetum (sheet of fibers of corpus callosum) forms the roof and lateral wall. The posteriorly sweeping optic radiation is separated with tapetum from the cavity of the posterior horn. The upper part of the medial wall is formed by the forceps major (fibers of the occipital lobe sweeping backward). The calcar avis, the lower part of the medial wall corresponds to the in-folding of the anterior part of the calcarine sulcus. There is no choroid plexus at the anterior and posterior horn.

3.1.2.3 The inferior (temporal) horn

The inferior horn is located inside of the temporal lobe, and it is the longest and largest of the 3 horns [2]. It creates a trajectory around the posterior end of the thalamus, goes posterolaterally and anteriorly into the temporal lobe. The roof is covered laterally by the inferior surface of the tapetum of the corpus callosum and medially by the tail of the caudate nucleus and stria terminalis. The floor is formed medially by collateral eminence produced by the hippocampus and laterally by a collateral sulcus. The hippocampal fibers form the alveus which covers the ventricular surface and form the fimbria converges medially. On the most medial side on the floor, the choroid plexus passes through the choroid fissure rest. The choroid plexus passes from the lateral ventricle into the inferior horn. The amygdaloid complex is situated at the anterior end of the inferior horn [21, 22]. The atrium (collateral trigone) connects the body of the lateral ventricle with the occipital and temporal horns.

3.2 Foramen of monro

Interventricular foramen of Monro is the communication between lateral and third ventricles, and it is bordered by the fornix, caudate nucleus, septum pellucidum, corpus callosum, and thalamus. The size of the ventricles determined the size and shape of the foramen. If the ventricular size is big, each foramen is rounded-shaped. As the ventricular size decreases, the foramen takes a crescent shape. The medial posterior choroidal arteries, the septal veins, and the superior choroidal vein pass through this structure [23].

3.3 Third ventricle

The third ventricle is located in-between the 2 thalami and some portion of the hypothalamus. This narrow vertical cavity of the diencephalon communicates with the lateral ventricles in the anterosuperior aspect, while on its posteroinferior aspect through the cerebral aqueduct of Sylvius, it communicates with the fourth ventricle [2]. The third ventricular cavity is lined by ependyma and is traversed by massa intermedia (interthalamic adhesion) that connects the 2 thalami which are located posterior to the foramen of Monro. It has a roof, a floor, 2 lateral walls, anterior and posterior walls.

A sheet of ependyma forms the **roof** which connects the upper border of the lateral wall of the ventricle. A triangular fold of pia mater, tela choroidea, covers the roof and it gives rise to the choroid plexus of the third ventricle. The **floor** descends ventrally and is formed by the optic chiasma, mammillary body, tuber cinereum, infundibulum, posterior perforated substance, and tegmentum of the midbrain [2].

From the interventricular foramen to the cerebral aqueduct, a curved hypothalamic sulcus extends and forms the **lateral wall**. The lateral wall is divided into 2 parts by the sulcus. The medial surface of the anterior two-thirds of the thalamus forms the larger upper part. The hypothalamus forms the smaller lower part and it is continuous with the floor. The anterior columns of the fornix that divided laterally into the lateral walls, the anterior commissure, and the lamina terminalis form the **anterior wall**. The lamina terminalis is a thin sheet of gray matter that extends superiorly from the rostrum of the corpus callosum, inferiorly to the optic chiasma. The cerebral aqueduct, the pineal gland, and the posterior commissure form the **posterior wall**.

The anterior recess (vulva of the ventricle), infundibular recess, optic recess, pineal recess, and supraspinal recess are the protrusions of the third ventricle into surrounding structures [24].

3.4 The aqueduct of sylvius

The Sylvian aqueduct measures 18 mm approximately and it is the narrowest part of the brain ventricular system. From the second fetal month, the luminal size of the aqueduct reduces due to the development of neighboring neural tissue [25]. The interventricular blockade mostly occurs here.

3.5 Fourth ventricle

The fourth ventricle is a wide, diamond-shaped cavity of the hindbrain [2]. It is located posterior to the pons and rostral part of the medulla, and anteroinferior to the cerebellum. On the sagittal section, it is seen as triangular, and on the horizontal section, it is seen as a rhomboidal shape. The floor of the fourth ventricle is also named the rhomboid fossa. It is continuous superiorly with the cerebral aqueduct and inferiorly with the central canal of the spinal cord. The fourth ventricle has 2 *lateral recesses*, a *medial dorsal recess*, and 2 *lateral dorsal recess*.

The fourth ventricle is bounded superolateral by the superior cerebellar peduncle and inferolateral by cuneate and gracile tubercles and inferior cerebellar peduncles.

Two superior cerebellar peduncles form the cephalic portion of the roof. Their medial margins overlap the ventricle on reaching the inferior colliculi. Superior medullary velum bridges the space between the superior cerebellar peduncle. Dorsally, it is covered by the lingula of the superior vermis of the cerebellum. The caudal portion of the roof is covered by the inferior medullary velum, which is formed by the tela choroidea of the fourth ventricle and the ventricular ependyma.

The lateral foramen of Luschka (located near the flocculus of the cerebellum) and the median foramen of Magendie (a large midline aperture, located in the roof of the ventricle at the lower part of inferior medullary velum) are the openings where the fourth ventricle communicates with the subarachnoid space. Mostly, the CSF passes through the medial foramen into the cerebellomedullary cistern, i.e., cisterna magna. The cerebral aqueduct does not contain choroid plexus.

4. The histology and embryology and the ventricular system

Ependymocytes (ependyma), which are a special type of cells that are columnar or cuboidal epithelium derived from the neuroepithelium cover the ventricular system of the brain [2]. The choroid plexus lies just below the ependymal layer and is responsible for CSF production.

A layer of subependymal glial cells tighten with the astrocyte processes and form the blood-brain barrier. Circumventricular organs are lack this barrier and have fenestrated capillaries with increased permeability. They have secretory and sensory functions. These are the area postrema, median eminence, pineal gland, organum vasculum of lamina terminalis, neurohypophysis, subcommissural organs, and sub-fornical organ [26]. The ciliary movement is oriented in the anteroposterior neuroaxis which is essential for the movement of CSF.

The ventricular system of the brain develops from the cavity of the neural tube [2]. Around the fourth week of gestation the neural tube is formed. Soon after, the spinal neurocele closes, and the neural cavity is separated from the amniotic cavity.

The choroid plexuses firstly appear in the 4th ventricle on the 41st day [27]. Different embryonic tissues give rise to cerebral and spinal meninges. At the third month of intrauterine life, the three meningeal layers differentiate [1]. The choroid plexus epithelium which is derived from the neural tube is continuous with the ependyma. The leptomeningeal axis is derived from the paraxial mesoderm. From the 26th week, cerebral veins dilate in the superior sagittal sinus at their anastomosis site. In the 35th week, the arachnoid villi are formed. The arachnoid stroma lined by endothelium protrudes into the lumen of the superior sagittal sinus via a defect in the dura mater. At the 39th week, real arachnoid granulations appear [28] and continue to develop around 18 months [29].

5. The blood supply and lymphatics of the ventricular system

The choroid plexus of the lateral ventricle is supplied from the anterior and posterior choroidal arteries, which are the internal carotid artery and the posterior cerebral artery branches respectively [2]. The posterior choroidal arteries supply the choroid plexus of the third ventricle. The anterior and posterior inferior cerebellar arteries supply the choroid plexus of the fourth ventricle.

6. The physiology of the cerebrospinal fluid

6.1 The cerebrospinal fluid secretion

Normal CSF formation rate is about 0.35 ml/min for adults, and this ranges from 400 to 600 ml per day [1, 30, 31]. CSF is renewed about four times a day. CSF production is elevated nocturnally and this may be due to cerebral metabolism alterations during sleep [32]. CSF formation also alters in disease states [33]. The choroid plexuses of the lateral ventricles and the tela choroidea of the third and fourth ventricles are responsible for most of the CSF secretion (60–70%) [1, 30]. Other sources of CSF are interstitial fluid, ependyma, and capillaries.

An asymmetrically positioned ion transporters at the blood- and CSF-facing membranes mediates fluid secretion into the ventricles. The choroid plexus epithelium is like the kidney proximal tubule, and transfer copious volumes of fluid [34]. The net transfer of sodium (Na^+) and chloride (Cl^-) from blood to ventricles determine CSF production [35–37]. From plasma across the basolateral membrane, Na^+ entry into choroid plexus epithelium is on a downhill gradient. Potassium (K^+), Cl^-, and bicarbonate (HCO_3^-) move downhill across the apical membrane into CSF at the other side of the choroid cell. These downhill ionic movements are set up by uphill active transport through the primary Na^+ pump both basolaterally and apically. This process requires chemical energy as adenosine triphosphate [ATP]. Choroid cell Na^+ concentration is kept relatively low by active Na^+ pumping into CSF [38] so a basolateral inward driving force for Na^+ transport from plasma into the epithelium was established [39].

For fluid formation the epithelial transport polarity is essential. From blood to CSF net fluid movement is enabled by the polar distribution of certain active transporters and passive channels. Streaming of ions and water were mediated by basolateral (interstitial) and apical (CSF) transporters and channels.

The direction of fluxes in CSF formation is mostly from interstitium to parenchyma to ventricles. K^+ and Cl^- passive diffusion (apical efflux) were allowed by channels into nascent CSF [40]. CSF formation is mainly produced by net secretion of Na^+, Cl^-, and HCO_3^-. Other ions, i.e., K^+, Mg^{2+}, and Ca^{2+} also have a role. Through the apical membrane water osmotically follows ion transport.

6.1.1 Sodium secretion

In CSF formation, the pivotal initiating step is the primary active transport of Na^+ from choroidal epithelium to ventricle [41]. Na^+, K^+-ATPase creates the electrochemical gradient, generates ATP, and empowers Na^+ pumping [42]. While CSF is being produced, the apical Na^+ efflux is balanced by permanent basolateral Na^+ influx through the epithelial Na^+ channel (ENaC) and Na^+-inward transport coupled with HCO_3^-, by the Na^+, HCO_3^- cotransporter, NBCn2/NCBE in order to equilibrate choroid pH and epithelial volume [39, 43–45].

6.1.2 Chloride secretion

Chloride, one of the primary anion in CSF secretion, is actively transported through the transcellular route in exchange for cellular HCO_3^- across the basolateral membrane [46]. Then for gathering above electrochemical equilibrium, plasma Cl^- goes into the epithelium [47]. In certain circumstances, intraepithelial Cl^- diffuses into CSF through the efflux arm of the Na^+-K^+-Cl^- cotransporter [48]. The downhill diffusion of Cl^- into CSF across apical Cl^- channels is the main pathway by which Cl^- accesses the ventricles to sustain fluid formation [40].

6.1.3 Bicarbonate secretion

In the choroid plexus, HCO_3^- has two sources. First, in choroid plexus epithelial cells to form H^+ and HCO_3^- ions carbonic anhydrase catalyzes the hydration of carbon dioxide (CO_2) [49]. Acetazolamide inhibits CSF secretion at least 50% which indicates that carbonic anhydrase is involved in CSF secretion [50]. Additionally, via Na^+-coupled HCO_3^- transport, HCO_3^- is pulled from plasma into the epithelium [43]. When the HCO_3^- accumulates, by two mechanisms it is ready for release through the CSF facing membrane. Firstly, in the epithelium downhill through an anion channel HCO_3^- diffuses into CSF [51]. Secondly, at the apical membrane HCO_3^- is transferred through an electrogenic Na^+-coupled HCO_3^- cotransporter [45, 52]. CSF rich in HCO_3^- show increased movement of HCO_3^- into ventricles as CSF is produced [53].

6.1.4 K^+ transport

K^+ enters into the cells in two ways: from the blood by the Na^+-K^+-$2Cl^-$ cotransporter-1 (NKCC1) and from the interstitial fluid by the Na^+-pump [54]. On both sides the influx exceeds the net flux across the cells, and thus at each membrane, there are thought to be pathways for efflux of most of the K^+ that enters through K^+ channels.

6.1.5 Water secretion

CSF has excretory, distributive, and buffering functions [31]. Water constitutes 99% of the CSF. After osmotically active Na^+, Cl^-, and HCO_3^- ions are transport into CSF, water follows them into the ventricles via a transcellular route by diffusing down its chemical potential gradient in the apical membrane through aquaporin 1 (AQP1) channels [55, 56]. AQP1 facilitates water transport from the interstitium to the CSF in the luminal and basolateral membrane [42]. Across the choroid plexus, transcellular water diffusion is a potential drug target to modulate CSF dynamics [55]. In regulating water molecule traffic through AQP1 channels agents structurally related to acetazolamide, furosemide, and bumetanide, and steroid hormones show promise [55, 57]. The composition of CSF and comparison with serum content were summarized in **Table 1**.

6.2 Regulation of cerebrospinal fluid regulation secretion and composition

The choroid plexuses receive adrenergic, peptidergic, cholinergic, and serotoninergic autonomic innervation [1]. The cholinergic system increases CSF secretion while the sympathetic nervous system reduces CSF secretion. Circadian variations of CSF secretion may be regulated by the autonomic nervous system.

The targets of humoral regulation are enzymes and membrane transporters. The activity of carbonic anhydrase is regulated by the acid-base disorders, membrane carrier proteins (i.e., the NaK_2Cl cotransporter), and aquaporins. Neuropeptide factors and monoamines also have a role. Atrial Natriuretic Peptide (ANP), Arginine Vasopressin (AVP) dopamine, serotonin, and melatonin receptors are present on the surface of the choroidal epithelium. ANP and AVP decrease CSF secretion [60], as ANP acts on AQP1.

Carbonic anhydrase inhibitors and loop diuretics decrease CSF secretion and turnover via enzymatic mechanisms, which could change the neuronal milieu, making prone the elderly to age-related neurodegenerative disorders.

6.3 Cerebrospinal fluid circulation

There is a one-way rostrocaudal CSF flow in ventricular cavities and a multi-way CSF flow in subarachnoid spaces from the sites of secretion to the sites of absorption (**Figure 2**) [1]. CSF flow is mainly affected by the systolic pulse wave in choroidal arteries and rapid respiratory waves. In the lateral ventricles, through interventricular foramina, CSF enters the third ventricle, and through the cerebral aqueduct, it enters the fourth ventricle. Thereafter, through the foramen of Magendie CSF goes to the subarachnoid spaces. Rostrally CSF circulates to the

Substance	CSF	Serum
Water content (% wt)	99	93
Total protein (mg/dl)	35	7000
Glucose (mg/dl)	60	90
Osmolarity (mOsm/l)	295	295
Sodium (mmol/l)	138	140
Potassium (mmol/l)	2.8	4.0
Calcium (mmol/l)	2.1	4.8
Magnesium (mEq/l)	2.0–2.5	1.7
Chloride (mmol/l)	119	103
pH	7.33	7.41

Table 1.
The composition of cerebrospinal fluid and comparison with serum [58, 59].

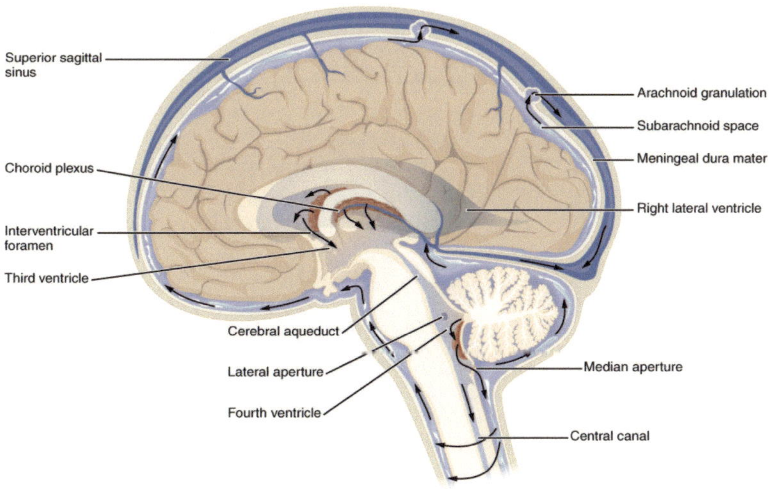

Figure 2.
Important structures for cerebrospinal fluid circulation [61].

villous sites of absorption and caudally it circulates to the spinal subarachnoid space in the cranial subarachnoid space. The spinal arachnoid villi partly absorb the CSF, and CSF circulates rostrally to the cranial subarachnoid space.

The subcommissural organ also has a role in CSF circulation. It is a differentiation of the ependyma at the rostral extremity of the cerebral aqueduct and synthesizes SCO-spondin [62]. This protein accumulates and forms Reissner fibers, which direct the CSF circulation via the cerebral aqueduct. Early during development in man, the subcommissural organ disappears. Certain forms of congenital hydrocephalus could be explained by an intrauterine abnormality of the subcommissural organ [63].

6.4 Cerebrospinal fluid absorption

CSF circulation was determined mainly by the arterial pulse from secretion site to absorption site [1]. The main site for CSF absorption into the venous outflow system are the cranial and spinal arachnoid villi. The cribriform plate, the cranial and spinal nerve sheaths, and the adventitia of cerebral arteries may also serve as alternative pathways for CSF drainage into the lymphatic system.

Arachnoid villi or granulations are endothelium-lined finger-like protrusions of the arachnoid outer layer via the dura mater in the venous sinus lumen (**Figure 3**) [65]. Villous absorption of CSF both in the brain or spine is a dynamic process that adapts the filtration rate to CSF pressure. The pressure gradient among the venous sinus and subarachnoid spaces is essential to assure CSF drainage is between 3 and 5 mmHg [66]. Especially during physical exertion, spinal arachnoid villi and the epidural venous plexus offer an alternative pathway for CSF absorption.

The cranial and spinal nerve sheaths and ependyma can also absorb CSF with respect to pressure gradients [1]. Via Virchow-Robin perivascular spaces, absorption through the interstitial compartment happens. On meningeal sheaths, CSF absorption surfaces have also been shown, especially the meningeal recesses of cranial and spinal nerve roots (i.e., the trigeminal and cochlear nerve). In its meningeal sheath, the optic nerve exerts a long extracranial course. With constructive interference in steady-state magnetic resonance imaging, a high-intensity ring around the optic nerve was observed in hydrocephalus. This finding indicates that when needed there is also a salvage pathway for reabsorption.

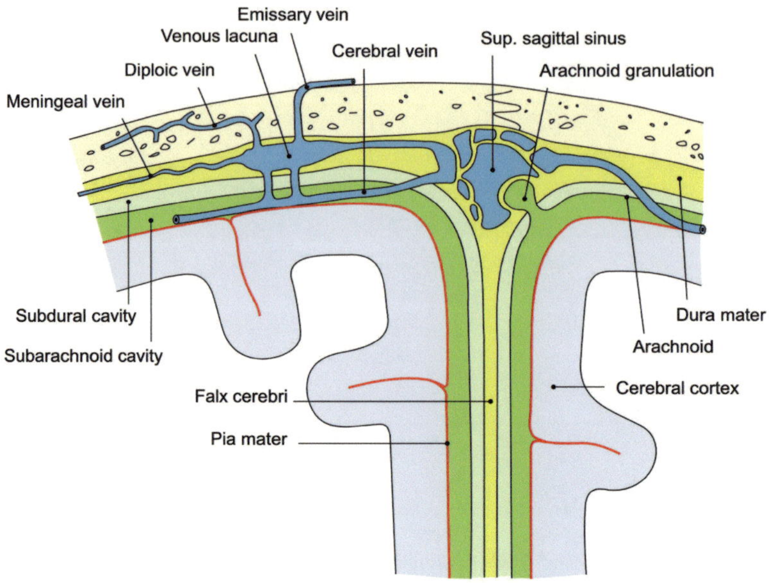

Figure 3.
Diagrammatic representation of a section across the top of the skull, showing the membranes of the brain, and arachnoid granulation [64].

When the cranial arachnoid villi capacities are exceeded, lymphatic absorption of CSF establishes an accessory pathway [1]. Particularly, this pathway is active in neonates. Arachnoid villi are completely functional after the age of 18 months. It becomes dysfunctional in the elderly due to fibrous changes of arachnoid granulations.

The cochlear aqueduct, which is located in the petrous part of the temporal bone, has a connection between the perilymphatic space of the cochlea and the subarachnoid space of the posterior cranial fossa. It is patent in 93% of cases [67]. This could clarify the effect of intracranial pressure changes on cochlear function (i.e., tinnitus after ventriculoperitoneal shunting and at high altitude).

6.5 Cerebrospinal fluid pressure

Ventricular cavity is a dynamic pressure system. CSF pressure is defined as the intracranial pressure (ICP) in the prone position. It is the outcome of a dynamic equilibrium between CSF secretion, resistance to flow, and absorption [1]. CSF pressure can be monitored invasively with a pressure transducer placed in the brain parenchyma or via an external ventricular/lumbar drain connected to CSF spaces. On Doppler ultrasound, to evaluate CSF pressure vascular flow can be traced as a non-invasive method. CSF pressure determines ICP with physiological values. In infancy, it ranges between 3 and 4 mmHg, and in adults, it ranges between 10 and 15 mmHg. Higher values indicate intracranial hypertension. Respiratory waves, jugular venous pressure, state of arousal, abdominal pressure, the subject's posture, and physical effort also modulate CSF flow dynamics and pressure.

The cranial content includes parenchymal, venous, and CSF compartments. CSF pressure is established by parenchymal and venous pressures. When fontanelles are open if ICP increases macrocephaly could be observed due to an increase in intracranial volume. When the fontanelles are closed, blood volume (particularly venous) reduction is seen as compensation.

Brain compliance is described as the volume needed to change ICP. It is the indication of the intracranial contents capacity to adapt to volume changes. Brain compliance is lower in men and changes with age. The volume needed to induce a

10-times increase in ICP is 8 ml in neonates, 20 ml in 2-year-old children, and 26 ml in adults. The brain volume must be considered when brain compliance is calculated (average of 335 ml in neonates and 1250 ml in young adults).

The regulation of CSF pressure occurs at the secretion, circulation, absorption phase of CSF. When intraventricular pressure is increased, the cerebral perfusion pressure (CPP) and the pressure gradient across the blood-CSF barrier decreases, and choroidal secretion is negatively affected. The concentrations of neuropeptides (ANP and AVP) in CSF and their receptor expression in the choroidal epithelium increase with CSF pressure increase and in the state of acute hydrocephalus [68, 69]. These neuropeptides cause a decreased choroidal secretion of CSF. They also induce dilatation of pial arteries to compensate for the reduction of CPP in acute hydrocephalus [70].

6.6 Cerebrospinal fluid homeostasis

CSF protects the neuraxis hydromechanically. CSF plays an important role in the regulation of cerebral interstitial fluid and the neuronal environment via arranging the circulation of active molecules, electrolyte balance, and elimination of catabolites. Via CSF, the products of choroid plexus secretion are transported to their action sites. The activity of certain brain regions is modulated by impregnation by this way. However, more rapid changes of activities happen via synaptic transmission [71].

7. Hydrocephalus

In hydrocephalus, an increased amount of fluid accumulates in the brain ventricular system [2]. Impairment in the CSF circulation at any point could lead to this disease. Mostly, abnormal enlargement of the cerebral ventricle and increased ICP are observed. The common symptoms include headache, irritability, blurred vision, vomiting, gait disturbance, and drowsiness. A rapid increase in head circumference is the main sign in infants.

Hydrocephalus can be classified as communicating or non-communicating type. Impaired absorption of CSF by the arachnoid granulations causes communicating hydrocephalus and this can be the result of any leptomeningeal processes (i.e., inflammation due to infectious or carcinomatous meningitis or hemorrhage as in acute subarachnoid hemorrhage).

Hydrocephalus can also be classified as congenital or acquired. Aqueductal stenosis is the most common cause of congenital hydrocephalus. This can be seen in the case of aqueductal atresia (genetical) or in the case of tumors of neighboring structures compressing the aqueduct or epididymitis (acquired). This results in the enlargement of both lateral and third ventricles with a normal fourth ventricle. The foramen of Magendie and Luschka could be obstructed in Chiari malformation. In this condition, the downward displacement of the cerebellum via the foramen magnum could result in internal hydrocephalus. In the case of inflammatory fibrosis of the meninges, the foramen could be obstructed and result in congenital hydrocephalus [25]. Trauma, infection, tumor, and hemorrhage could result in acquired hydrocephalus.

8. Surgical approaches

Here we will briefly mention the most preferred methods for hydrocephalus management.

In ventriculostomy, a hole in the ventricles is created for CSF drainage and/or ICP monitoring. External ventricular drain is placed in the ventricle. Kocher's point is the commonest entry point on the skull which is 3–4 cm lateral to the midline and 11 cm posterior to the glabella. The frontal horn of the lateral ventricle is the target [72].

In ventricular shunting, CSF is diverted from ventricles to body compartments such as the peritoneal cavity (ventriculoperitoneal shunt), right atrium (ventriculoatrial shunt), pleural space (ventriculopleural shunt).

In order to drain the CSF directly into the basal cisterns, an incision could also be made on the floor of the third ventricle.

9. Conclusion

CSF is an essential liquid for brain homeostasis. It has a well-balanced ionic content and has a certain secretion and absorption rate. Choroid plexuses are the main secretion site, while arachnoid villi are the main absorption site. When there is disequilibrium in secretion, absorption or any obstruction in the ventricular system hydrocephalus could be seen. There are alternative treatment methods for this condition which depend on the etiology. In this chapter, we review the historical understanding of ventricular anatomy and CSF, main anatomical structures of the ventricular system, histology of embryology of the ventricular system, CSF physiology. We also briefly mentioned hydrocephalus and the main treatment alternatives.

Conflict of interest

The authors declare no conflict of interest.

Appendices and nomenclature

ANP	atrial natriuretic peptide
AQP1	aquaporin 1
AVP	arginine vasopressin
BC	before Christ
Cl^-	chloride
CO_2	carbondioxide
CPP	cerebral perfusion pressure
CSF	cerebrospinal fluid
ENaC	epithelial Na^+ channel
HCO_3^-	bicarbonate
ICP	intracranial pressure
Na^+	sodium
NKCC1	Na-K-2Cl cotransporter-1
Wt	weight

Author details

Pinar Kuru Bektaşoğlu[1]* and Bora Gürer[2]

1 Department of Neurosurgery, University of Health Sciences, Istanbul Fatih Sultan Mehmet Education and Research Hospital, Istanbul, Turkey

2 Department of Neurosurgery, Istinye University Faculty of Medicine, Istanbul, Turkey

*Address all correspondence to: drpinarkuru@gmail.com

References

[1] Sakka L, Coll G, Chazal J. Anatomy and physiology of cerebrospinal fluid. European Annals of Otorhinolaryngo logy, Head and Neck Diseases. 2011; **128**(6):309-316. DOI: 10.1016/j.anorl. 2011.03.002

[2] Shenoy SS, Lui F. Neuroanatomy, ventricular system. Updated 2020 July 31. In: StatPearls. Treasure Island (FL): StatPearls Publishing; 2021

[3] Hippocrates. Collected Works (Translated and Edited by Jones WHS). London: W. Heinemann; 1923

[4] Marshall LH, Magoun HW. Discoveries in the human brain. In: Neuroscience Prehistory, Brain Structure, and Function. New Jersey: Humana Press; 1998

[5] Romero R. Herophilos, the great anatomist of antiquity. Anatomy. 2015;**9**(2):108-111. DOI: 10.2399/ ana.15.003

[6] Pearce JM. The neuroanatomy of herophilus. European Neurology. 2013;**69**(5):292-295. Available from: http://doi.org/b47s

[7] Duque-Parra JE, Barco-Ríos J, García-Aguirre JF. A historical approach to the ventricular system of the brain. Revista de la Facultad de Medicina. 2017;**65**(3):473-477. DOI: 10.15446/ revfacmed.v65n3.57884

[8] Kühn K, Editor. Claudii Galeni Opera Omnia (Cambridge Library Collection - Classics). Cambridge: Cambridge University Press. 2011. DOI: 10.1017/ CBO9780511895289

[9] Rengachary SS, Colen C, Dass K. Development of anatomic science in the late middle ages: The roles played by Mondino de Liuzzi and Guido da Vigevano. Neurosurgery. 2009;**65**(4):787-793. Available from: http://doi.org/b9nkg2

[10] Gross CG. Brain, vision, memory. In: Tales in the History of Neuroscience. Cambridge: The MIT Press; 1999

[11] Swedenborg E. The Brain, Considered Anatomically, Physiologically and Phylosopically (Translated and Edited by Tafel RL). London: James Speirs; 1882-1887

[12] von Haller A. Primae lineae physiologiae in usum praelectionum academicarum. Gottingae: Vandenhoeck; 1747

[13] Schiller F. The cerebral ventricles. From soul to sink. Archives of Neurology. 1997;**54**(9):1158-1162. DOI: 10.1001/archneur.1997.00550210086018

[14] Hajdu SI. A note from history: discovery of the cerebrospinal fluid. Clinical & Laboratory Science. 2003 Summer;**33**(3):334-336

[15] Willis T. Cerebri Anatome cui Accessit Nervorum Descriptio et Usus. London: J Martyn & J Allestry; 1664

[16] Wynter WE. Four cases of tubercular meningitis in which paracentesis of the theca vertebralis was performed for the relief of fluid pressure. Lancet. 1891;**1**:981

[17] Quincke H. Die Lumbal punction des Hydrocephalus. Berliner Klinische Wochenschrift. 1891;**28**:929-933, 965-968

[18] Mestrezat W. Le liquide céphalo-rachidien normal et pathologique, valeur clinique de l'examen chimiqe. Paris: Maloine; 1912

[19] Cushing HW. Studies on the cerebrospinal fluid. Journal of Medical Research. 1914;**8**:406-409

[20] Case courtesy of Dr. Matt Skalski from the case rID: 37808. Available from: https://radiopaedia.org/cases/37808

[21] FitzGerald MJT, Folan-Curran J. Clinical Neuroanatomy and Related Neuroscience. 4th ed. Philadelphia, Pa: WB Saunders; 2002

[22] Waxman SG. Ventricles and coverings of the brain. In: Correlative Neuroanatomy. 24th ed. New York, NY: Lange Medical Books/McGraw-Hill; 2000. pp. 153-168

[23] Tubbs RS, Oakes P, Maran IS, Salib C, Loukas M. The foramen of Monro: A review of its anatomy, history, pathology, and surgery. Child's Nervous System. 2014;**30**(10):1645-1649. DOI: 10.1007/s00381-014-2512-6

[24] Krokfors G, Katila O, Taalas J. Enlarged suprapineal recess of the third ventricle. Acta Neurologica Scandinavica. 1967;**43**(5):607-615

[25] Bickers DS, Adams RD. Hereditary stenosis of the aqueduct of Sylvius as a cause of congenital hydrocephalus. Brain. 1949;**72**(Pt. 2):246-262

[26] Ganong WF. Circumventricular organs: Definition and role in the regulation of endocrine and autonomic function. Clinical and Experimental Pharmacology & Physiology. 2000;**27**(5-6):422-427. DOI: 10.1046/j.1440-1681.2000.03259.x

[27] O'Rahilly R, Müller F. The meninges in human development. Journal of Neuropathology and Experimental Neurology. 1986;**45**:588-608

[28] Gómez DG, DiBenedetto AT, Pavese AM, Firpo A, Hershan DB, Potts DG. Development of arachnoid villi and granulations in man. Acta Anatomica. 1981;**111**:247-258. DOI: 10.1159/000145473

[29] Chazal J, Tanguy A, Irthum B, Janny P, Vanneuville G. Dilatation of the subarachnoid pericerebral space and absorption of cerebrospinal fluid in the infant. Anatomia Clinica. 1985;7:61-66

[30] Johanson CE, Duncan JA 3rd, Klinge PM, Brinker T, Stopa EG, Silverberg GD. Multiplicity of cerebrospinal fluid functions: New challenges in health and disease. Cerebrospinal Fluid Research. 2008;**5**:10. DOI: 10.1186/1743-8454-5-10

[31] Johanson C. Choroid plexus-CSF circulatory dynamics: Impact on brain growth, metabolism and repair. In: Conn P, editor. Neuroscience in Medicine. Totowa, New Jersey: The Humana Press; 2008. pp. 173-200

[32] Nilsson C, Ståhlberg F, Gideon P, Thomsen C, Henriksen O. The nocturnal increase in human cerebrospinal fluid production is inhibited by a beta 1-receptor antagonist. The American Journal of Physiology. 1994;**267**(6 Pt 2): R1445-R1448. DOI: 10.1152/ajpregu.1994.267.6.R1445

[33] Silverberg GD, Heit G, Huhn S, Jaffe RA, Chang SD, Bronte-Stewart H, et al. The cerebrospinal fluid production rate is reduced in dementia of the Alzheimer's type. Neurology. 2001;**57**(10):1763-1766. DOI: 10.1212/wnl.57.10.1763

[34] Spector R, Snodgrass SR, Johanson CE. A balanced view of the cerebrospinal fluid composition and functions: Focus on adult humans. Experimental Neurology. 2015;**273**:57-68. DOI: 10.1016/j.expneurol.2015.07.027

[35] Knuckey NW, Fowler AG, Johanson CE, Nashold JR, Epstein MH. Cisterna magna microdialysis of 22Na to evaluate ion transport and cerebrospinal fluid dynamics. Journal of Neurosurgery. 1991;**74**(6):965-971. DOI: 10.3171/jns.1991.74.6.0965

[36] Johanson CE, Palm DE, Dyas ML, Knuckey NW. Microdialysis analysis of effects of loop diuretics and acetazolamide on chloride transport from blood to CSF. Brain Research. 1994;**641**(1):121-126. DOI: 10.1016/0006-8993(94)91823-6

[37] Smith QR, Woodbury DM, Johanson CE. Kinetic analysis of [36Cl]-, [22Na]- and [3H]mannitol uptake into the in vivo choroid plexus-cerebrospinal fluid brain system: Ontogeny of the blood brain and blood-CSF barriers. Brain Research. 1982;**255**(2):181-198. DOI: 10.1016/0165-3806(82)90019-0

[38] Smith QR, Johanson CE. Effect of ouabain and potassium on ion concentrations in the choroidal epithelium. The American Journal of Physiology. 1980;**238**(5):F399-F406. DOI: 10.1152/ajprenal.1980.238.5.F399

[39] Murphy VA, Johanson CE. Na(+)-H+ exchange in choroid plexus and CSF in acute metabolic acidosis or alkalosis. The American Journal of Physiology. 1990;**258**(6 Pt 2): F1528-F1537. DOI: 10.1152/ajprenal.1990.258.6.F1528

[40] Brown PD, Davies SL, Speake T, Millar ID. Molecular mechanisms of cerebrospinal fluid production. Neuroscience. 2004;**129**(4):957-970. DOI: 10.1016/j.neuroscience.2004.07.003

[41] Johanson CE, Reed DJ, Woodbury DM. Active transport of sodium and potassium by the choroid plexus of the rat. The Journal of Physiology. 1974;**241**(2):359-372. DOI: 10.1113/jphysiol.1974.sp010660

[42] Damkier HH, Brown PD, Praetorius J. Cerebrospinal fluid secretion by the choroid plexus. Physiological Reviews. 2013;**93**(4):1847-1892. DOI: 10.1152/physrev.00004.2013

[43] Praetorius J. Water and solute secretion by the choroid plexus. Pflügers Archiv. 2007;**454**(1):1-18. DOI: 10.1007/s00424-006-0170-6

[44] Chen LM, Kelly ML, Rojas JD, Parker MD, Gill HS, Davis BA, et al. Use of a new polyclonal antibody to study the distribution and glycosylation of the sodium-coupled bicarbonate transporter NCBE in rodent brain. Neuroscience. 2008;**151**(2):374-385. DOI: 10.1016/j.neuroscience.2007.10.015

[45] Johanson CE, Murphy VA. Acetazolamide and insulin alter choroid plexus epithelial cell [Na+], pH, and volume. The American Journal of Physiology. 1990;**258**(6 Pt 2): F1538-F1546. DOI: 10.1152/ajprenal.1990.258.6.F1538

[46] Lindsey AE, Schneider K, Simmons DM, Baron R, Lee BS, Kopito RR. Functional expression and subcellular localization of an anion exchanger cloned from choroid plexus. Proceedings of the National Academy of Sciences of the United States of America. 1990;**87**(14):5278-5282. DOI: 10.1073/pnas.87.14.5278

[47] Smith QR, Johanson CE. Active transport of chloride by lateral ventricle choroid plexus of the rat. The American Journal of Physiology. 1985;**249**(4 Pt 2): F470-F477. DOI: 10.1152/ajprenal.1985.249.4.F470

[48] Keep RF, Xiang J, Betz AL. Potassium cotransport at the rat choroid plexus. The American Journal of Physiology. 1994;**267**(6 Pt 1): C1616-C1622. DOI: 10.1152/ajpcell.1994.267.6.C1616

[49] Johanson CE. Differential effects of acetazolamide, benzolamide and systemic acidosis on hydrogen and bicarbonate gradients across the apical and basolateral membranes of the choroid plexus. The Journal of

Pharmacology and Experimental Therapeutics. 1984;**231**(3):502-511

[50] Bradbury MWB. The Concept of a Blood–Brain Barrier. Chichester: Wiley; 1979

[51] Millar ID, JIe B, Brown PD. Ion channel diversity, channel expression and function in the choroid plexuses. Cerebrospinal Fluid Research. 2007;**4**:8. DOI: 10.1186/1743-8454-4-8

[52] Millar ID, Brown PD. NBCe$_2$ exhibits a 3 $HCO_3(-)$:1 Na+ stoichiometry in mouse choroid plexus epithelial cells. Biochemical and Biophysical Research Communications. 2008;**373**(4):550-554. DOI: 10.1016/j.bbrc.2008.06.053

[53] Husted RF, Reed DJ. Regulation of cerebrospinal fluid bicarbonate by the cat choroid plexus. The Journal of Physiology. 1977;**267**(2):411-428. DOI: 10.1113/jphysiol.1977.sp011820

[54] Hladky SB, Barrand MA. Fluid and ion transfer across the blood-brain and blood-cerebrospinal fluid barriers; a comparative account of mechanisms and roles. Fluids Barriers CNS. 2016;**13**(1):19. DOI: 10.1186/s12987-016-0040-3

[55] Johanson CE. Fluid-Forming Functions of the Choroid Plexus: What Is the Role of Aquaporin-1? 1st ed. Enfield, New Hampshire: Science Publishers; 2015. pp. 140-171

[56] Johansson PA, Dziegielewska KM, Ek CJ, Habgood MD, Møllgård K, Potter A, et al. Aquaporin-1 in the choroid plexuses of developing mammalian brain. Cell and Tissue Research. 2005;**322**(3):353-364. DOI: 10.1007/s00441-005-1120-x

[57] Huber VJ, Tsujita M, Nakada T. Aquaporins in drug discovery and pharmacotherapy. Molecular Aspects of Medicine. 2012;**33**(5-6):691-703. DOI: 10.1016/j.mam.2012.01.002

[58] Irani DN. Cerebrospinal Fluid in Clinical Practice. Michigan: Elsevier Health Sciences; 2008. ISBN 9781416029083

[59] Watson MA, Scott MG. Clinical utility of biochemical analysis of cerebrospinal fluid. Clinical Chemistry. 1995;**41**(3):343-360. Erratum in: Clinical Chemistry. 1995;**41**(8 Pt 1): 1207

[60] Faraci FM, Mayhan WG, Heistad DD. Effect of vasopressin on production of cerebrospinal fluid: Possible role of vasopressin (V1)-receptors. The American Journal of Physiology. 1990;**258**(1 Pt 2): R94-R98. DOI: 10.1152/ajpregu.1990.258.1.R94

[61] Bruce Blaus. Blausen.com staff. Medical gallery of Blausen Medical 2014. WikiJournal of Medicine. 2014;**1**(2). DOI: 10.15347/wjm/2014.010. ISSN 2002-4436. Own work

[62] Gobron S, Creveaux I, Meiniel R, Didier R, Dastugue B, Meiniel A. SCO-spondin is evolutionarily conserved in the central nervous system of the chordate phylum. Neuroscience. 1999;**88**(2):655-664. DOI: 10.1016/s0306-4522(98)00252-8

[63] Galarza M. Evidence of the subcommissural organ in humans and its association with hydrocephalus. Neurosurgical Review. 2002;**25**(4):205-215. DOI: 10.1007/s10143-002-0208-y

[64] OpenStax. OpenStax Anatomy and Physiology. 2016. Available fom: https://cnx.org/contents/FPtK1zmh@8.25:fEI3C8Ot@10/Preface. Version 8.25

[65] Welch K, Friedman V. The cerebrospinal fluid valves. Brain. 1960;**83**:454-469. DOI: 10.1093/brain/83.3.454

[66] Pollay M. The function and structure of the cerebrospinal fluid

outflow system. Cerebrospinal Fluid Research. 2010;**7**:9. DOI: 10.1186/1743-8454-7-9

[67] Gopen Q, Rosowski JJ, Merchant SN. Anatomy of the normal human cochlear aqueduct with functional implications. Hearing Research. 1997;**107**:9-22. DOI: 10.1016/s0378-5955(97)00017-8

[68] Yamasaki H, Sugino M, Ohsawa N. Possible regulation of intracranial pressure by human atrial natriuretic peptide in cerebrospinal fluid. European Neurology. 1997;**38**(2):88-93. DOI: 10.1159/000113166

[69] Mori K, Tsutsumi K, Kurihara M, Kawaguchi T, Niwa M. Alteration of atrial natriuretic peptide receptors in the choroid plexus of rats with induced or congenital hydrocephalus. Child's Nervous System. 1990;**6**(4):190-193. DOI: 10.1007/BF01850969

[70] Tamaki K, Saku Y, Ogata J. Effects of angiotensin and atrial natriuretic peptide on the cerebral circulation. Journal of Cerebral Blood Flow and Metabolism. 1992;**12**(2):318-325. DOI: 10.1038/jcbfm.1992.44

[71] Veening JG, Barendregt HP. The regulation of brain states by neuroactive substances distributed via the cerebrospinal fluid; a review. Cerebrospinal Fluid Research. 2010;**7**:1. DOI: 10.1186/1743-8454-7-1

[72] Munakomi S, Das JM. Ventriculostomy. In: StatPearls. Treasure Island (FL): StatPearls Publishing; 2021

Chapter 2

Circulation of Cerebrospinal Fluid (CSF)

Hayriye Soytürk, Murat Yılmaz, Cansu Önal, Eylem Suveren and Ümit Kılıç

Abstract

Circulation of cerebrospinal fluid (CSF) is a clear, colorless liquid that circulates between the ventricular system and the subarachnoid space. In addition to its function as a natural cushion for the brain, CSF provides the circulation of metabolic products, hormones, and neurotransmitters. Moreover, it has tasks such as maintaining the homeostatic balance of the central nervous system, protecting the brain against mechanical injuries, preventing direct contact of the brain with the extracellular region. It also has a role in maintaining cerebral interstitial fluid (ISF) homeostasis and neuronal regulation. Normal CSF production, its circulation, and absorption have a critical role for the development and functioning of the brain. In an average adult person, roughly 150 ml of CSF circulates at any given moment. The ventricular part accounts for about 17% of the total volume of fluid, with the rest located in the subarachnoid cisterns and space. CSF is produced at a rate of about 0.3–0.4 mL/min, translating to 18–25 mL/H and 430–530 mL/day.

Keywords: circulation of cerebrospinal fluid (CSF), production of CSF, absorption of CSF, cerebral interstitial fluid (ISF), physiology of circulation of CSF

1. Introduction

A large amount of cerebrospinal fluid (CSF) is produced in the choroid plexus in the lateral ventricle. This produced CSF reaches the 3rd ventricle via the foramen Monroe, from here to the 4th ventricle via the Aquaduct Sylvius, and passes to the subarachnoid space via the foramen Magendie and the foramen Luschkas. From here, the CSF moves upwards, reaching the interpedincular cistern through the prepontin cistern, and from here to the convexity through the chiasmatic cistern. The CSF circulating from the dorsal, however, reaches convexity via the quadrigeminal cistern, ambient cistern, and vena cerebri magna cistern through the cerebellar hemispheres. CSF also passes into the central canal through the spinal cord and into the spinal subarachnoid space [1]. Absorption occurs from the arachnoid villi to the venous sinuses [2, 3].

The choroid plexus consists of villi. Each villi is covered with monolayer cubic epithelium with cilia. These cells are located on a basal membrane consisting of collagen, fibroblast, and nerve fibers. In the middle of each villi is a capillary vessel with no tight connections between endothelial cells with a loose wall structure. The blood–brain barrier in the choroid plexus is not formed by the tight connections between endothelial cells in the capillaries, unlike the parenchyma, but by the tight connections between cells on the villi [4].

Tight junction on the apical surface of the epithelium above the basal membrane form the blood–brain barrier [2, 4]. The choroid plexus, the epandymal layer, and the parenchyma are the production sites of CSF. As a result of studying isolated choroid plexus preparations, it has been shown that the choroid plexus is responsible for the production of approximately 80% of the CSF. However, the extrachoroidal source of CSF is not well known [5]. The first step of choroidal secretion involves passive filtration of plasma from the choroidal capillaries along a pressure gradient to the choroidal interstitial compartment.

CSF production from the choroid plexus occurs as a two-step process [6]. The first stage in CSF production is the accumulation of ultrafiltrate in the villi, which leaks out of the vein as a result of hydrostatic pressure due to tight connections between endothelial cells from the capillaries in the middle of the villi. This accumulation is secreted by the choroid plexus cell by converting it into CSF. Ultrafiltrate, which accumulates on the side of the basal membrane, is transferred by the sodium-potassium pump which actively pumps sodium into the cell, while water passively enters the cell along with sodium. The same pump also helps chloride. In addition, chloride enters the cell independently of this pump. Fluid with CSF properties inside the cell is again actively released into the ventricular cavity by the cell wall facing the ventricle with the help of a sodium-potassium pump [4, 5]. In normal physiological conditions, CSF production is not affected by intracranial pressure, but when intracranial pressure increases, there may be a decrease in CSF production, as it will affect ultrafiltrate production, which is the first stage of CSF production [5]. Changes in body temperature and serum osmolarity are not effective in CSF production [6].

CSF absorption occurs in arachnoid granulations located along the superior sagittal sinus. Coming from the subarachnoid region to the venous lakes in the arachnoid granulations, the CSF is absorbed into the cerebral veins. It is believed that these pathways work more, especially during an increase in intracranial pressure. CSF absorption is sensitive to pressure. CSF absorption increases when intracranial pressure increases, and absorption decreases if it is below the basal value [4, 7]. This is followed by the bulk ion transport from blood to CSF which occurs via the transcellular pathway facilitated by membrane ion carrier proteins and cytoplasmic carbonic anhydrase. Carbonic anhydrases catalyze the conversion of H_2O and CO_2 to H^+ and HCO_3. Transcellular transport of sodium (Na^+) and chloride (Cl^-) (together with HCO_3) are the most important ions carried by the choroid plexus epithelium, forming the osmotic gradient that activates H_2O secretion [8].

ATP-dependent ion pumps of the apical membrane allow the passage of Na^+, Cl^-, HCO_3, and Potassium (K^+) ions towards the ventricular lumen, forming the electrochemical gradient for CSF secretion. Transepithelial water movement also occurs through the transcellular pathway. It follows the osmotic gradient created by ATP dependent mechanisms and is facilitated by the aquaporins (AQP1) of the apical basolateral and luminal membranes [8, 9].

Aquaporins, which have different variations in the body, are found in many tissues. It has tissue specific variants. They are expressed in various secretory epithelium. For example, AQP1 and AQP5 are expressed in the pancreas, AQP3 and AQP5 are expressed in the salivary glands. AQP1 is abundantly expressed in the choroid plexus epithelium and contributes to the high water permeability of the membrane [8].

There are some substances that affect CSF production, and one of these substances is furosemide. Feurosemide's mechanism has little effect on carbonic anhydrase, but its main effect is that it can reduce CSF production by stopping chloride entering the cell. Another substance is acetazolamide, which causes a decrease in CSF production in humans and experimental models by blocking the carbonic anhydrase enzyme in the cell [5].

In physiological conditions, the rate of CSF production should be equal to the rate of absorption. Given that production and absorption occur in different parts of the system, it is assumed that its flow rate will be affected [9].

Oreskovic and Klarica studied the effects of the choroid in CSF physiology via plexectomies [9]. According to classical theory, a large decrease in the overall secretion of CSF is expected when choroid plexectomy is performed, and therefore some pressure relief can be achieved in patients with hydrocephalus with this method. Studies have shown that two-thirds of patients treated for a recurrence of hydrocephalus should be shunted [10].

The choroid plexectomy study conducted by Oreskovic and Klarica on rhesus monkeys showed that the chemical component of CSF remained normal, suggesting that the choroid plexus played a less role in molecular transport [11]. While there is evidence to support these claims regarding CSF production, there is also a large amount of literature describing the tides and the net flow of CSF [12].

According to Spector, active secretion and absorption of CSF are carried out by movable cilia located in the ependymal wall. CSF involves the transport of growth factors to certain areas of the brain in the circulatory system [12]. The composition of cerebrospinal fluid, CSF, consists mainly of 99% of water, while the remaining 1% consists of proteins, ions, neurotransmitters, and glucose [8, 13, 14]. The concentration, total viscosity, and surface tension of each of these proteins found in CSF vary in different conditions [15, 16], CSF absorption increases, and if below the basal value, absorption decreases [4, 17].

This is followed by the bulk ion transport from blood to CSF which occurs via the transcellular pathway facilitated by membrane ion carrier proteins and cytoplasmic carbonic anhydrase. Carbonic anhydrases catalyze the conversion of H_2O and CO_2 to H^+ and HCO_3. Transcellular transport of Na^+ and Cl^- (together with HCO_3) are the most important ions carried by the choroid plexus epithelium, forming the osmotic gradient that activates H_2O secretion [8].

The composition of CSF differs from serum due to different expressions of membrane-associated channels and transport proteins, which in this case causes the choroidal epithelium to be unidirectional [18]. On the apical side, epithelial cells are interconnected by tight junction that limit the movement of these molecules, and intercellular space connections form the blood–brain barrier. The apical side of the epithelium is covered with microvilli, while the basolateral side has folds that increase the surface area of the cells and make it more suitable for absorption.

Compared to plasma, CSF usually contain high concentrations of sodium (Na^+), chloride (Cl^-), and magnesium (Mg^{+2}), while lower concentrations contain potassium (K^-) and calcium (Ca^{+2}) [14]. On the apical side, active transport pumps release ions into the ventricular cavities. The movement of water in the apical membrane has been shown to be caused by the presence of aquaporin 1 (AQP1). Indeed, a study by Mobasheri and Marples revealed that the choroid plexus is among the tissues with the highest expression of AQP1 in the body [19].

Many studies have different conclusions regarding the AQP 4 as the main candidate for water transport in the basolateral membrane. The common focus of the studies has been the study of disease conditions that affect the production, absorption, or composition of CSF. Apart from its mechanical role, CSF has an important role in biochemical homeostasis throughout the CNS [12].

By using new techniques to analyze the diversity of CSF components, proteins, lipids, hormones and microRNAs, it will be possible to track the development of the disease over time in disease conditions [20]. The production and absorption of some CSF biomolecules, such as growth factors, neurotransmitters, cytokines, extracellular matrix proteins, permeability-related proteins, binding proteins, and adhesion molecules can affect CSF homeostasis. Similarly, the microenvironment surrounding periventricular cells and their activities may vary in disease states [20].

CSF production is regulated by the autonomic nervous system and neuropeptides such as dopamine and atrial natriuretic peptide. The sympathetic nervous system reduces CSF production, while the cholinergic system increases its production. There is a circadian rhythm in CSF production [21]. Most of the hormones that regulate systemic water and electrolyte homeostasis, such as aldosterone, angiotensin II, and arginine vasopressin, are also present in the choroid plexus and ventricular system. These hormones are believed to have two tasks: the first is the production of CSF locally, and the second is the regulation of extracellular fluid in the brain, but they also have tasks in the central regulation of blood pressure [8]. CSF production can be reduced by the administration of diuretics and carbonic anhydrase inhibitors. In addition, any increase in intraventricular pressure can reduce plasma filtration and, as a result, CSF production by lowering the pressure gradient in the blood–brain barrier. In CSF and interstitial brain fluid, water and solutes change constantly, and this balance provides an optimal environment for neurons. This is directly proportional to the rate of formation of CSF and inversely proportional to the volume of CSF. In aging, there is less efficient active transport, with a slower CSF cycle causing the accumulation of potentially harmful metabolites in the interstitium of the brain. Clearance of brain metabolites per minute depends on the CSF regeneration at a rate of 0.3–0.4%. Brain catabolites form when fluid turnover rates drop by more than 50%, and the reduction in amyloid b decrease in CSF clearance is now believed to be associated with the development of Alzheimer's disease [22].

After production, CSF movement is usually carried out through the ventricular system, while it is also supported by the cilia ependyma [23]. The net flow of the CSF passes through the ventricular system, starting from the lateral ventricles [24]. The CSF flows from the lateral ventricles, through the left and right foramen of the Monro to the third ventricle. Then, it passes to the 4th ventricles. From the fourth ventricle, the CSF may flow laterally from the foramen of Lushka, or medially from the foramen of Magendie to the subarachnoid space. Passing through the foramen of Magendie results in the filling of the spinal subarachnoid space. CSF outflow from the foramen of Luschka goes into the subarachnoid space of cisterns and into the subarachnoid space that covers the cerebral cortex. CSF from the subarachnoid space is eventually reabsorbed into the superior sagittal sinus (SSS), known as the arachnoid. Arachnoid granulations provide reabsorption of CSF into the bloodstream by a pressure-dependent gradient [6]. In arachnoid granulations, outlets towards the CNS are seen due to the fact that the pressure in the subarachnoid space is greater than the venous sinus pressure. Similar to new theories about CSF production, there are also absorption theories. Studies in animal models have revealed that CSF can also be significantly absorbed through cervical lymphatics [6].

CSF, which is not reabsorbed by arachnoid granulations, can reach cervical lymphatics in two alternative ways. The first is along the subarachnoid space of the emerging cranial nerves [6]. This provides a direct route through which CSF can be transferred from cisterns to extracranial lymphatics. The second way in which CSF can reach lymphatics is through the Virchow-Robin space of the arteries and veins that penetrate the parenchyma of the brain [25].

The Virchow-Robin Space (VRS) is the area surrounding the arteries and veins of the brain parenchyma, which can vary in size depending on disease status. When the CSF is not absorbed by the classical way, it can enter the VRS or be directed to the brain interstitial fluid (ISF). The brain interstitial fluid ISF is believed to be a compartment with a subarachnoid space (SAS) that is mediated by AQP s and bidirectional flow to VRS, but it is not yet clear. If the CSF enters the ISF, it will either be reabsorbed into the bloodstream, or it will enter the VRS, or it will enter the subarachnoid space again. From the VRS, CSF can reenter into the SAS or be reabsorbed by cervical lymphatics, depending on the forces exerted by cardiac pulsations and pulmonary respiration. In addition to the circulation of CSF to cervical lymphatics, studies have also been conducted explaining the reabsorption of CSF to the dural venous plexus. Arachnoid granulations at birth are not fully developed, and CSF absorption is based on the venous plexus of the inner surface of dura, which is more robust in infants [26]. Although not common in adults, the dural venous plexus is believed to play a role in absorption. Adult and fetal cadaver dissections and animal models with intradural injection have all been shown to fill the parasagittal dural venous plexus [27].

2. Physiology of circulation of cerebrospinal fluid

The CSF physiology, in the classical sense, is based mainly on animal experiments [28]. In recent research, the structure of CSF circulation has been questioned, challenging significant aspects of the classical model. Recently, CSF production and absorption have been reevaluated [9, 29–31].

According to the classical view described by Cushing in 1926 as the "third circulation" [32, 33], CSF flows from the ventricular system through the Lushka and Magendie foramen into the subaracanoid area in a one-way, rostrocaudal manner. The CSF then continues to flow either downwards around the spinal cord or upwards over the cerebral convexities, and is eventually absorbed by arachnoid granulations and arachnoidal villi on either side of the upper sagittal sinus.

Recent studies have highlighted a secondary pathway of CSF, circulation through perivascular VRS, similar to the lymphatic system in other parts of the body [34, 35]. This CSF circulatory system, which has a similar function to the lymphatic system with the participation of astroglia, has been called the "glymphatic system" [36, 37]. The glial membrane of the brain consists of the astrocytic end-feet and forms theVRS, it has high amounts of aquaporin channels and facilitates CSF transfer from VRS to the interstitial space of the brain cavity is cleaned and then empty the drainage paths paravenous makes it easy to carry along [36]. The in vivo imaging taken using fluorescent substances in mice also showed how this microcirculation removes amyloid beta and other waste products from the central nervous system [34].

CSF flow is pulsatile and depends on pulsational arterial perfusion. A central ventricular pulse wave is formed, followed by brain expansion, followed by a subarachnoid CSF frontooccipital pulse wave [38]. During systole, blood flows into the brain, expanding into the brain, compressing the ventricles and the cortical vessels outwards and SAS. Inward expansion of the brain leads to the pulsatile transfer of CSF from the the cerebral aqueduct and the rest of the ventricular system. During diastole, the volume of the brain decreases, and CSF flows in the opposite direction along the the cerebral aqueduct and the foramen magnum. The movement of CSF to the brain through VRS is also supported by arterial vibrations [35]. This suggests a link between decreased arterial pulses, which are often seen in some elderly patients, and amyloid B accumulation in Alzheimer's disease [36, 37]. Although in-vivo studies in humans are needed to confirm these findings, there is growing evidence that plaque may be another key site for extracranial output [24, 39].

Since Cushing, the collective flow character of CSF circulation has been accepted by most researchers. Even in recent studies, it is assumed that the CSF circulation is directed towards the arachnoid villus along the ventricles and subarachnoid space [24, 40]. VRS are a histologically defined anatomical area surrounding blood vessels as they enter the brain tissue from the subarachnoid space Initially, VRSs were believed to be connected to the subarachnoid space, allowing for free fluid transfer. This concept was later elucidated by microscopic investigations that showed perivascular cavities as dead ends, open to the subarachnoid space but closed to the parenchyma, and therefore not a channel for flow [41].

Considering the microscopic anatomy of VRS, its thin structure is actually located on layers of endothelial, pial, and glial cells, each defined by different basal membranes [42]. The glia covering the brain parenchyma forms the outer wall of the VRSs [43]. In the capillary bed, the basal membrane of the glia merges with the outer vascular membrane, forming the VRS [44].

The arterial and venous vessels, which are located in the cortical subarachnoid space (SAS), are covered by a layer of pial cells that surround the vessels. The pial sheath forms a cavity next to the vessel wall, called the perivascular space (PVS) [45]. At the entrance of the cortical vessels to the VRS, the pial sheaths merge with the layer of pial cells lining the brain surface, forming a funnel-like structure that accompanies the VRS to the vessels only for a short distance [46]. However, the pial sheath of the arterial vessels extends to the VRS. Near the capillary bed, the pial sheath becomes more and more windowed and leaky [45].

Some authors use the terms "Virchow-Robin space" and "perivascular space" as synonyms [47], while others use the terms to name different areas as discussed above [48]. Studies with electron microscopy show that pial membranes separate VRS from the cortical subarachnoid space [46]. Since electron microscopy of human brain samples shows that VRS and PVS have collapsed, it has been a matter of debate whether these histologically characterized compartments are really openings or spaces [45]. However, studies in rodents have shown that VRS is filled with fluid, electron microscopic dense material [46], macrophages and other inflammatory cells [42].

Although pial cell layers separate the VRS from the cortical subarachnoid space, physiologically there is strong evidence that fluid circulates throughout the VRS. There are species-related differences in the pial layer. In mice, for example, the pial layer is very thin, while in humans it is thicker [49].

In humans, the pial sheath is described as a sensitive but seemingly continuous layer of cells, connected by desmosomes and cavity connections but without obvious tight connections [50]. As a result of numerous experimental studies, it has been recognized that the pia mater does not have permeable properties against liquids [51]. Given that the flow within the VRS depends on the pulsatility of the arteries [52], hydrostatic forces can move liquids and solutes along the pial membranes. However, while VRS basically allows for a two-way exchange between CSF and ISF, there is not much data to explain the scope and kinetics of such fluid movements.

Although it has been shown that the pial membranes between PVS and SAS can prevent the exchange of larger molecules, the intraparenchymal injection has not been shown to spread to cisternal CSF, although it has accumulated in PVS [53]. This observation is supported by clinical findings that red blood cells are confined in the subarachnoid space and do not enter the VRS following aneurysm rupture in humans [49]. It has been shown both experimentally and clinically that PVS, and possibly, more importantly, pathways between the essential membranes of arterioles and the wall of arteries, provide drainage for ISF and the brain's waste molecules.

There is experimental evidence that paraarterial drainage pathways are connected to the lymphatics of the posterior skull base [54]. In reality, the solutes and fluids can be discharged through the VRS from the brain interstitium through the arteries, into the cervical lymphatics [55]. This view was supported experimentally by immunohistochemical and confocal microscopic observations showing that fluorescent dyes such as 3 kD dextran or 40 kD ovalbumin move along the basic membranes of capillaries and arteries after being injected into the corpus striatum in mice.

These findings are clinically significant as beta-amyloid accumulates in the vascular wall of arterioles and arteries, based on observations in patients with cerebral amyloid angiopathy. The accumulation of insoluble amyloid can block this drainage pathway and therefore inhibit the elimination of beta-amyloid and interstitial fluid from the brain in Alzheimer's disease [54]. The size of amyloid deposition is so pronounced that it has been proposed as a natural determinant for peri-arterial drainage pathways [55]. Peri-arterial drainage of liquids and solutes has important effects not only in neurodegenerative diseases but also in immunological CNS diseases [55]. Similar to arteries, veins in the subarachnoid space have pial sheath forming a PVS [42].

Author details

Hayriye Soytürk[1]*, Murat Yılmaz[2], Cansu Önal[3], Eylem Suveren[4] and Ümit Kılıç[5]

1 Department of Interdisciplinary Neuroscience, Bolu Abant Izzet Baysal University, Bolu, Turkey

2 Department of Neurology, Bolu Abant Izzet Baysal University, Bolu, Turkey

3 Department of Biology, Bolu Abant Izzet Baysal University, Bolu, Turkey

4 Departmant of Nursing, Bolu Abant Izzet Baysal University, Bolu, Turkey

5 Department of Physiology, Düzce University, Düzce, Turkey

*Address all correspondence to: hayriyesoyturk1@gmail.com

References

[1] Sadler, T. W. (1996). Langman'S Medikal Embriyoloji (7. Baskı). C. Başaklar, K. Sönmez. Ankara: Palme Yayıncılık.

[2] Sklar, F. (1996). Physiology of the cerebrospinal fluid compartment. Neurosurgery. 2nd ed. New York: McGraw-Hill, 3, 3617-3623.

[3] Sato, O., Oi, S., & Yamada, S. (1999). Hydrocephalus: experimental considerations and clinical analyses. Choux M, Di Rocco C, Hockley AD. Pediatric Neurosurgery, 11-237.

[4] Rekate, H. (1990). Current concepts of CSF production and absorption. Hydrocephalus, 11-22.

[5] Detwiler, P. W., Porter, R. W., & Rekate, L. H. (1999). Hydrocephalus-clinical features and management. Choux M, Di Rocco C, Hockley AD. Pediatric Neurosurgery, 12-253.

[6] Brinker, T., Stopa, E., Morrison, J., & Klinge, P. (2014). A new look at cerebrospinal fluid circulation. Fluids and Barriers of the CNS, 11(1), 1-16.

[7] Milhorat, T. H. (1996). Hydrocephalus: Pathophysiology and clinical features. Neurosurgery, 3625-3631.

[8] Damkier, H. H., Brown, P. D., & Praetorius, J. (2013). Cerebrospinal fluid secretion by the choroid plexus. Physiological reviews, 93(4), 1847-1892.

[9] Orešković, D., Klarica, M., & Vukić, M. (2002). The formation and circulation of cerebrospinal fluid inside the cat brain ventricles: a fact or an illusion?. Neuroscience letters, 327(2), 103-106.

[10] Lapras, C., Mertens, P., Guilburd, J. N., Pialat, J., & Patet, J. D. (1988). Choroid plexectomy for the treatment of chronic infected hydrocephalus. Child's Nervous System, 4(3), 139-142.

[11] Orešković, D., & Klarica, M. (2010). The formation of cerebrospinal fluid: nearly a hundred years of interpretations and misinterpretations. Brain research reviews, 64(2), 241-262.

[12] Spector, R., Snodgrass, S. R., & Johanson, C. E. (2015). A balanced view of the cerebrospinal fluid composition and functions: focus on adult humans. Experimental neurology, 273, 57-68.

[13] Bulat, M., & Klarica, M. (2011). Recent insights into a new hydrodynamics of the cerebrospinal fluid. Brain research reviews, 65(2), 99-112.

[14] Sakka, L., Coll, G., & Chazal, J. (2011). Anatomy and physiology of cerebrospinal fluid. European annals of otorhinolaryngology, head and neck diseases, 128(6), 309-316.

[15] Brydon, H. L., Hayward, R., Harkness, W., & Bayston, R. (1995). Physical properties of cerebrospinal fluid of relevance to shunt function. 1: The effect of protein upon CSF viscosity. British journal of neurosurgery, 9(5), 639-644.

[16] Brydon, H. L., Hayward, R., Harkness, W., & Bayston, R. (1996). Does the cerebrospinal fluid protein concentration increase the risk of shunt complications?. British journal of neurosurgery, 10(3), 267-274.

[17] Milhorat, T. H. (1996). Hydrocephalus: Pathophysiology and clinical features. Neurosurgery, 3625-3631.

[18] Brown, P. D., Davies, S. L., Speake, T., & Millar, I. D. (2004). Molecular mechanisms of cerebrospinal fluid production. Neuroscience, 129(4), 955-968.

[19] Mobasheri, A., & Marples, D. (2004). Expression of the AQP-1 water channel in normal human tissues: a semiquantitative study using tissue microarray technology. American Journal of Physiology-Cell Physiology.

[20] Johanson, C., & Johanson, N. (2016). Merging transport data for choroid plexus with blood-brain barrier to model CNS homeostasis and disease more effectively. CNS & Neurological Disorders-Drug Targets (Formerly Current Drug Targets-CNS & Neurological Disorders), 15(9), 1151-1180.

[21] Nilsson, C., Stahlberg, F., Thomsen, C., Henriksen, O., Herning, M., & Owman, C. (1992). Circadian variation in human cerebrospinal fluid production measured by magnetic resonance imaging. American Journal of Physiology-Regulatory, Integrative and Comparative Physiology, 262(1), R20-R24.

[22] Silverberg, G. D., Mayo, M., Saul, T., Rubenstein, E., & McGuire, D. (2003). Alzheimer's disease, normal-pressure hydrocephalus, and senescent changes in CSF circulatory physiology: a hypothesis. The Lancet Neurology, 2(8), 506-511.

[23] Roales-Buján, R., Páez, P., Guerra, M., Rodríguez, S., Vío, K., Ho-Plagaro, A., & Jiménez, A. J. (2012). Astrocytes acquire morphological and functional characteristics of ependymal cells following disruption of ependyma in hydrocephalus. Acta neuropathologica, 124(4), 531-546.

[24] Johanson, C. E., Duncan, J. A., Klinge, P. M., Brinker, T., Stopa, E. G., & Silverberg, G. D. (2008). Multiplicity of cerebrospinal fluid functions: new challenges in health and disease. Cerebrospinal fluid research, 5(1), 1-32.

[25] Cherian, I., Beltran, M., Kasper, E. M., Bhattarai, B., Munokami, S., & Grasso, G. (2016). Exploring the Virchow–Robin spaces function: A unified theory of brain diseases. Surgical neurology international, 7(Suppl 26), S711.

[26] Mack, J., Squier, W., & Eastman, J. T. (2009). Anatomy and development of the meninges: implications for subdural collections and CSF circulation. Pediatric radiology, 39(3), 200-210.

[27] Papaiconomou, C., Zakharov, A., Azizi, N., Djenic, J., & Johnston, M. (2004). Reassessment of the pathways responsible for cerebrospinal fluid absorption in the neonate. Child's Nervous System, 20(1), 29-36.

[28] Hassin, G. B. (1947). The Cerebrospinal Fluid Pathways: A Critical Note. Journal of Neuropathology & Experimental Neurology, 6(2), 172-176.

[29] Bateman, G. A., & Brown, K. M. (2012). The measurement of CSF flow through the aqueduct in normal and hydrocephalic children: from where does it come, to where does it go?. Child's Nervous System, 28(1), 55-63.

[30] Greitz, D. (1993). Cerebrospinal fluid circulation and associated intracranial dynamics. A radiologic investigation using MR imaging and radionuclide cisternography. Acta radiologica. Supplementum, 386, 1-23.

[31] Bulat, M., Lupret, V., Orešković, D., & Klarica, M. (2008). Transventricular and transpial absorption of cerebrospinal fluid into cerebral microvessels. Collegium antropologicum, 32(1), 43-50.

[32] Cushing, H. (1926). Studies in Intracranial Physiology & Surgery: The Third Circulation: the Hypophysis: the Gliomas. Oxford University Press.

[33] Black, P. M. (1999). Harvey cushing at the Peter Bent Brigham hospital. Neurosurgery, 45(5), 990-1001.

[34] Iliff, J. J., Wang, M., Liao, Y., Plogg, B. A., Peng, W., Gundersen, G. A., Nedergaard, M. (2012). A paravascular pathway facilitates CSF flow through the brain parenchyma and the clearance of interstitial solutes, including amyloid β. Science translational medicine, 4(147), 147ra111.

[35] Iliff, J. J., Lee, H., Yu, M., Feng, T., Logan, J., Nedergaard, M., & Benveniste, H. (2013). Brain-wide pathway for waste clearance captured by contrast-enhanced MRI. The Journal of clinical investigation, 123(3), 1299-1309.

[36] Rasmussen, M. K., Mestre, H., & Nedergaard, M. (2018). The glymphatic pathway in neurological disorders. The Lancet Neurology, 17(11), 1016-1024.

[37] Jessen, N. A., Munk, A. S. F., Lundgaard, I., & Nedergaard, M. (2015). The glymphatic system: a beginner's guide. Neurochemical research, 40(12), 2583-2599.

[38] Preuss, M., Hoffmann, K. T., Reiss-Zimmermann, M., Hirsch, W., Merkenschlager, A., Meixensberger, J., & Dengl, M. (2013). Updated physiology and pathophysiology of CSF circulation—the pulsatile vector theory. Child's Nervous System, 29(10), 1811-1825.

[39] Johnston, M., Zakharov, A., Papaiconomou, C., Salmasi, G., & Armstrong, D. (2004). Evidence of connections between cerebrospinal fluid and nasal lymphatic vessels in humans, non-human primates and other mammalian species. Cerebrospinal fluid research, 1(1), 1-13.

[40] Pardridge, W. M. (2011). Drug transport in brain via the cerebrospinal fluid. Fluids and Barriers of the CNS, 8(1), 1-4.

[41] Woollam, D. H. M., & Millen, J. W. (1955). The perivascular spaces of the mammalian central nervous system and their relation to the perineuronal and subarachnoid spaces. Journal of anatomy, 89(Pt 2), 193.

[42] Krueger, M., & Bechmann, I. (2010). CNS pericytes: concepts, misconceptions, and a way out. Glia, 58(1), 1-10.

[43] Krahn, V. (1982). The pia mater at the site of the entry of blood vessels into the central nervous system. Anatomy and embryology, 164(2), 257-263.

[44] Bechmann, I., Priller, J., Kovac, A., Böntert, M., Wehner, T., Klett, F. F. & Nitsch, R. (2001). Immune surveillance of mouse brain perivascular spaces by blood-borne macrophages. European Journal of Neuroscience, 14(10), 1651-1658.

[45] Zhang, E. T., Inman, C. B., & Weller, R. O. (1990). Interrelationships of the pia mater and the perivascular (Virchow-Robin) spaces in the human cerebrum. Journal of anatomy, 170, 111.

[46] Krisch, B. (1988). Ultrastructure of the meninges at the site of penetration of veins through the dura mater, with particular reference to Pacchionian granulations. Cell and tissue research, 251(3), 621-631.

[47] Ichimura, T., Fraser, P. A., & Cserr, H. F. (1991). Distribution of extracellular tracers in perivascular spaces of the rat brain. Brain research, 545(1-2), 103-113.

[48] Thal, D. R. (2009). The Precapillary Segment of the Blood-Brain Barrier and Its Relation to Perivascular Drainage in Alzheimer's Disease and Small Vessel Disease. The Scientific World Journal, 9, 557-563.

[49] Hutchings, M., & Weller, R. O. (1986). Anatomical relationships of the pia mater to cerebral blood vessels in man. Journal of neurosurgery, 65(3), 316-325.

[50] Alcolado, R., Weller, R. O., Parrish, E. P., & Garrod, D. (1988). The cranial arachnoid and pia mater in man: anatomical and ultrastructural observations. Neuropathology and applied neurobiology, 14(1), 1-17.

[51] Cserr, H. F., Depasquale, M., Patlak, C. S., & Pullen, R. G. (1986). Convection of cerebral interstitial fluid and its role in brain volume regulation. Annals of the New York Academy of Sciences, 481(1), 123-134.

[52] Hadaczek, P., Yamashita, Y., Mirek, H., Tamas, L., Bohn, M. C., Noble, C., & Bankiewicz, K. (2006). The "perivascular pump" driven by arterial pulsation is a powerful mechanism for the distribution of therapeutic molecules within the brain. Molecular Therapy, 14(1), 69-78.

[53] Szentistvanyi, I. S. T. V. A. N., Patlak, C. S., Ellis, R. A., & Cserr, H. F. (1984). Drainage of interstitial fluid from different regions of rat brain. American Journal of Physiology-Renal Physiology, 246(6), F835-F844.

[54] Weller, R. O., Djuanda, E., Yow, H. Y., & Carare, R. O. (2009). Lymphatic drainage of the brain and the pathophysiology of neurological disease. Acta neuropathologica, 117(1), 1.

[55] Carare, R. O., Bernardes-Silva, M., Newman, T. A., Page, A. M., Nicoll, J. A. R., Perry, V. H., & Weller, R. O. (2008). Solutes, but not cells, drain from the brain parenchyma along basement membranes of capillaries and arteries: significance for cerebral amyloid angiopathy and neuroimmunology. Neuropathology and applied neurobiology, 34(2), 131-144.

Section 2

Normal Pressure Hydrocephalus

Chapter 3

Normal Pressure Hydrocephalus: Revisiting the Hydrodynamics of the Brain

Fernando Hakim, Daniel Jaramillo-Velásquez, Martina González, Diego F. Gómez, Juan F. Ramón and Mateo Serrano-Pinzón

Abstract

Normal pressure hydrocephalus syndrome is the most common form of hydrocephalus in the elderly and produces a dementia which can be reversible surgically. It is characterized by ventriculomegaly and the classic triad of symmetric gait disturbance, cognitive decline and urinary incontinence, also known as Hakim's triad. To date, the exact etiology of the disease has not been elucidated and the only effective treatment is a cerebrospinal fluid shunting procedure which can be a ventriculoatrial, ventriculoperitoneal or lumboperitoneal shunt. The most important problem is the high rate of underdiagnosis or misdiagnosis due to similarities in symptoms with other neurodegenerative disorders, and in some cases, coexistence. Hence, increasing awareness amongst the community and medical professionals in order to increase clinical suspicion, timely diagnosis and treatment are paramount. The best way to achieve this is by having a structured protocol with patient-centered tests that evaluates the entire myriad of alterations a clinician might encounter whenever treating patients with this disorder. Recent advances in imaging technology as well as cerebrospinal fluid biomarkers have given interesting insight into the pathophysiology of the disease and will certainly contribute greatly in diagnostic advancements. We finally present an institutional protocol which has been accredited by international peers with promising results in diagnostic and outcome rates.

Keywords: Normal pressure hydrocephalus syndrome (NPH), cerebrospinal fluid (CSF), ventriculoatrial shunt (VAS), ventriculoperitoneal shunt (VPS), lumboperitoneal shunt, intracranial pressure (ICP), lumbar puncture (LP), center of excellence (COE)

1. Introduction

Normal Pressure Hydrocephalus (NPH) is a type of chronic communicating hydrocephalus recognized as the most common form of hydrocephalus in the elderly with an estimated prevalence of 5.9% in those aged over 80 years [1]. NPH was first described by Hakim in 1965, hence, it is sometimes found as Hakim's syndrome. NPH is characterized by enlarged brain lateral ventricles (ventriculomegaly) with normal cerebrospinal fluid (CSF) pressure and the classic clinical triad of symmetric gait disturbance, gradually progressive cognitive decline, and urinary incontinence, also known as Hakim's triad [2].

In this chapter, we present a concise review of all available evidence regarding epidemiology, pathophysiology, diagnostic and treatment methods including some diagnostic novelties that could be determinant sometime in the near future.

2. Epidemiology

NPH is the main cause of surgically reversible dementia in the population aged 50–80 years with a global estimated prevalence of 0.2% in the group aged 70–79 years and 5.9% in those over 80 according to epidemiologic studies in Sweden [1]. However, the estimated prevalence reported by populated-based studies in Japan is 1.6% [3–6]. If surveys in Sweden had used diagnostic criteria in Japanese guidelines, their weighted average would have been 1.5% [7]. These data may underestimate the actual prevalence and incidence because some studies have flawed methodology and use different criteria. The prevalence and incidence of any hydrocephalus, especially NPH, will increase as mean life expectancy increases. Mortality associated with untreated hydrocephalus ranges from 20 to 87% depending on the etiology [8]. Therefore, diagnosing and treating chronic hydrocephalus is a public health problem that requires attention and active participation of the entire scientific community and health personnel.

NPH is a condition with a high rate of underdiagnosis because it can be confused with other types of dementia, especially Alzheimer's dementia (AD) and Parkinson's disease, and because in many cases there is simply no clinical suspicion by medical staff. Approximately 5–10% of patients with some type of dementia may suffer from chronic hydrocephalus and another concomitant disorder such as AD, frontotemporal dementia (FTD), subcortical arteriosclerotic dementia (SAD) also known as Binswanger's disease, Lewy bodies dementia, amongst others. In fact in epidemiologic studies, 24% of patients with clinical suspicion of NPH had typical histopathologic findings of AD in brain tissue samples. Furthermore, the components of the clinical triad, although useful, are not always sufficient since they are not pathognomonic to NPH. Thirty-five percent (35%) of people over 70 have dementia from any cause, 40% of women and 20% of men over 60 have urinary incontinence from another cause, and 20% of people over 70 have some type of gait impediment [9].

To date, the only effective treatment for NPH is a CSF shunting procedure, typically ventriculo-peritoneal shunt (VPS) or ventriculo-atrial shunt (VAS) and sometimes, lumbo-peritoneal shunt (LPS), with a success rate that ranges between 60 and 80%. Hence the importance of appropriate diagnosis and adequate selection of patients for surgery.

3. Pathophysiology

The Monro-Kellie doctrine describes the relationship between the contents of the cranium and intracranial pressure (ICP). The doctrine dictates that the adult cranial vault is a rigid, non-expansible container inside of which there is constant balance of blood, CSF, and brain tissue. If one of the components increases, one of the others or both must decrease, or vice versa, to keep the ICP within the normal range (5–15 mm Hg). This system of homeostasis in the CNS is known as autoregulation. Although the brain is viscoelastic and can change shape and volume after mechanical stimuli, it is the blood and CSF that determine this balance. If one of the components increases in a progressive fashion, and overcomes the autoregulatory mechanisms, a point of no return or decompensation is reached with an increase in central venous pressure (CVP) and an exponential increase in ICP. When this happens, cerebral perfusion pressure (CPP), which is the difference between mean arterial pressure (MAP) and ICP (CPP = MAP-ICP) decreases, causing brain edema. If this is not corrected, intracranial hypertension (ICH) invariably ensues. It is considered pathologic when ICP levels are more than 30 mm Hg, and life-threatening if more than 40 mm Hg. [10]. CVP is therefore essential in the pathophysiology of ICH.

CSF is the electrochemical medium that provides the appropriate conditions for neuronal metabolism, in addition, since the CNS is immersed, its net weight is ~20–30 g, considerably reducing the risk of mechanical damage [9]. Eighty to ninety percent (80–90%) of CSF is produced in the choroid plexuses of the four cerebral ventricles, richly vascularized epithelial structures whose epithelium constitutes the blood-cerebrospinal fluid brain barrier (BCSFBB) [11]. The remaining 10–20% is produced in the extra-choroidal blood–brain barrier (BBB). Starling's laws (hydrostatic and oncotic pressure gradients), as well as the ionic concentration within the medium are probably the main mechanisms responsible for production, although the exact mechanism is still unknown [11]. CSF flows from the lateral ventricles through the foramen of Monro to the third ventricle, from there through the cerebral aqueduct or Sylvian aqueduct in a pulsatile manner, closely related to heartbeat and respiratory rate, to the fourth ventricle. From the fourth ventricle it passes through the medial opening of Magendie and lateral foramina of Luschka to the subarachnoid space (SAS) and flows upward bathing the tentorium towards the venous sinuses. A smaller amount of volume flows down the spinal SAS. Production is constant and so is reabsorption. Classically it has been said that CSF reabsorption occurs in arachnoid granulations and villi, fenestrated endothelial structures that protrude macroscopically and microscopically, respectively, towards the lumen of the venous sinuses, especially the superior sagittal sinus. It is collectively accepted that retrograde flow from arachnoid granulations is impossible. Some authors point out that almost a third of the CSF leaves the neural sheaths of the cranial nerves to enter the glymphatic system [11]. Therefore, the exact mechanism of reabsorption also remains to be elucidated.

ICP gradients cancel as long as there is constant CSF flow and thus there is no risk of brain herniation. Important concept because when there are abrupt changes in pressure gradients as happens in traumatic brain injury (TBI) or space-occupying lesions, the critical threshold is 20–25 mm Hg, whereas pressure-volume studies in patients with communicating hydrocephalus indicate that these patients tolerate values of ICP in the range of 40–50 mm Hg without symptoms of ICH [12](7). Now, the fact that NPH patients have ventriculomegaly and symptomatic hydrocephalus, but normal ICP is what constitutes the same clinical problem and paradoxical situation that interested S. Hakim and still eludes us today. However, it is important to highlight that some patients with NPH may have episodes of elevated ICP, especially in the early stages of the disease. This phenomenon was explained by S. Hakim based on Laplace's law that explains the behavior of a spherical elastic container containing fluid. The pressure of the fluid inside the container (P) is directly proportional to the elastic tension of its wall (T) and is inversely proportional to the radius (r) of the container [12]:

$$P = 2T/r \tag{1}$$

S. Hakim used the analogy of rubber balloons. As one inflates a balloon, initial pressure is higher than final pressure, as air volume inside the balloon increases, less pressure is required to be able to introduce more air. Therefore, due to the visco-elastic properties of the brain, as the size of the ventricles (the rubber balloon) increases, pressure required to stretch them decreases. In other words, the pressure inside the container increases until, after a certain stretch point, pressure drops with a subsequent increase in size independent of the wall tension. CSF pressure exerts greater force on the wall of dilated ventricles than in normal sized ventricles. Clarifying this concept, CSF pressure is the necessary pressure to prevent fluid from escaping through a needle that enters the SAS or the ventricle and represents the force per unit of ventricular surface area and/or subarachnoid surface. This is Pascals' law: the total force exerted (F) on the brain would be the result of the multiplication of the pressure (P) and the surface area of ventricular walls (A) subjected to the fluid pressure inside [2]:

$$F = P \times A \tag{2}$$

Hence,

$$P = F/A \tag{3}$$

To illustrate this, imagine that ICP (CSF pressure) is 150 mm H2O. This means that each squared millimeter of ventricular surface is subjected to the weight of a 150-mm column of water. Now imagine two ventricles of different sizes subjected to the same pressure:

Ventricle A (Av) with a surface area of 20 cm^2 and ventricle B (Bv) with a surface area of 40 cm^2. The force exerted on Av (FAv) is different from Bv (FBv).

$$FAv = 150 \times 20 \tag{4}$$

Whereas, in Bv:

$$FBv = 150 \times 40 \tag{5}$$

Thus, we now understand why pressure in both ventricles is the same, but the force exerted on the ventricular wall has increased in relation to the proportional increase in its surface area. Hakim called this phenomenon the *hydraulic press effect*. Of course, this analogy could be criticized, because the visco-elastic properties of the brain also depend on other variables, but due to its mathematical simplicity, it is pertinent [2].

Imbalance between CSF production and reabsorption is the cause of progressive ventriculomegaly. The reason why there is functional obstruction to CSF reabsorption is currently unknown. It must be an alteration in hydrodynamics. CSF reabsorption (Re), which is linear and unidirectional, depends on the gradient between the subarachnoid ICP (ICP_{SA}) and sagittal venous sinus pressure (P_{SS}), a constant parameter that depends on CVP, and is inversely proportional to CSF outflow resistance into the venous space (R_{CSF}), sometimes referred to as *Rout* calculated experimentally at 6–10 mm Hg mL-1 min^{-1} [12]:

$$Re = ICP_{SA} - P_{SS}/R_{CSF} \tag{6}$$

Venous compliance is the ability of veins to distend when intravascular volume and transmural pressure increases against tendency to return to the original dimensions on application of the distending force. In mechanics, it is the inverse of stiffness and the reciprocal of elastance or *coefficient of elasticity*. The increase in CSF within the ventricles causes ventricular dilation and mechanical stretching of the periventricular white matter, altering viscoelastic properties of the brain, compromising the Windkessel effect, widely studied in systemic circulation but its role in the CNS is also recognized. Mechanical stretching generates intraparenchymal capillary compression which eventually leads to ischemia of *watershed* areas in the territory of the anterior cerebral artery, which reasonably explains the symptoms of the clinical triad. Therefore, one of the most accepted theories about NPH etiology is decreased cortical venous compliance that increases R_{CSF} accompanied by chronic ischemia. This is the reason why if CSF pressure is reduced to subnormal levels after lumbar puncture (LP), venous pressure returns to normal and pushes the dilated ventricles back to their normal position, reversing hydrocephalus. Some authors have reported that $R_{CSF} > 12$ mm/Hg/mL/min is a suitable threshold for predicting responsiveness after surgery [13]. Parallelly, a rise of 3–4 mm Hg in P_{SS} can halt CSF absorption via the arachnoid granulations [14]. Experimental studies in hydrocephalus-induced animal models and prospective clinical studies are needed to elucidate the reasons why cortical venous compliance decreases and why R_{CSF} increases in NPH patients.

Chronic hypoperfusion have shown to cause progressive loss of compliance as well as astrogliosis and neuroinflammation. In fact, glial fibrillar acidic protein (GFAP), a marker of astrocyte axon reactivity and damage is increased in brain biopsy specimens of NPH patients. Additionally, astrogliosis further worsens parenchymal stiffness and contributes to decreased CSF hydrodynamics [15, 16]. Furthermore, NPH patients have significantly more blood–brain barrier (BBB) disruptions than controls which have shown to play a key role in neurologic dysfunction by allowing leakage and entry of blood-borne products into the CNS [17, 18]. BBB disruption and leakage are associated with greater astrogliosis [19].

The recently discovered glymphatic system, which presumably participates to a certain extent in CSF reabsorption, is mediated by aquaporin-4 (AQP4) channels on astrocytic perivascular pedicles [20]. Normal glymphatic system functioning depends on appropriate arterial pulsation, intact AQP4 and healthy sleep pattern [20]. NPH patients have a delayed removal of intrathecal tracers in phase-contrast imaging studies which suggests impaired functioning of the glymphatic system [21, 22]. Decreased AQP4 channel density in perivascular endfeet of NPH patients have also been reported. And it is not unusual that elderly patients suffer from concomitant sleep disorders that might contribute to the pathogenesis [23–27].

4. Etiology and classification

Hydrocephalus is a syndrome caused by a heterogeneous group of pathologies that share an increase in intracranial CSF volume manifested by ventriculomegaly and compatible symptoms. They can be classified as congenital or secondary (acquired), non-communicating or communicating, adult or childhood onset and with ICH or normal pressure. NPH is an acquired type of chronic communicating hydrocephalus with normal ICP whose etiology is still unknown.

NPH could be part of aging process and senescence of the venous endothelium and loss of brain viscoelastic properties, but it has not been possible to determine whether it is related to cerebrovascular and/or cardiovascular risk factors, since not all patients with NPH have a cardiocerebrovascular comorbidity. It could be a degenerative process secondary to the deposit of specific material such as beta-amyloid, something similar to what happens in AD, but its causal relationship has not been determined since not all patients with NPH have neurodegenerative comorbidity. The etiology of NPH still remains unknown for neurologists, neurosurgeons, and neurophysiologists. Thus, is known as idiopathic hydrocephalus.

5. Clinical manifestations

Unlike other types of hydrocephalus, neurologic exam of patients with NPH, except for the clinical triad and minor neurologic signs, is essentially normal. Important, because any sign that suggests focal deficit must make the clinician suspicious of other disorders. Paresis, hyperreflexia, and other first motor neuron signs are atypical. NPH should be suspected in adults with any of the three components of Hakim's triad, but it is not necessary for all three to be present to diagnose NPH. Of the triad components, typically the first one to appear, most frequently encountered and most severe is symmetric apraxia or magnetic gait not explained by another cause and for this reason some authors have called it the cardinal symptom [1]. It is also the first to resolve after surgery in patients with more than one component. Most reports agree that the triad is complete in approximately 60% of cases [28–30]. Nevertheless, a large-scale questionnaire conducted in Japan in 2012 revealed that only 12.1% of a cohort of 1524 patients had the full triad [31].

When all three components are present, the odds of NPH are higher. When there is cognitive impairment, it is important that the patient is accompanied by a family member, preferably his partner or one who lives with the patient to build a medical history and reach the correct clinical diagnosis. After detailed questioning, it has been confirmed that most patients have insidious onset of symptoms within a period of 3–6 months [1]. It is essential to inquire about past medical history, especially cranial surgery or intracranial bleeding, trauma, infection, or CNS tumors, since many of these patients are at risk of secondary hydrocephalus. In patients with ventriculomegaly who only have cognitive impairment or only have urinary incontinence, different diseases should be suspected.

Polyneuropathy is common in the elderly and is a frequent cause of urinary and motor comorbidity. Patients with gait impairment and incontinence without cognitive impairment should be studied for spinal lesions such as cervical or lumbosacral myelopathy secondary to cervical/lumbosacral canal stenosis or discovertebral disease at any level [1].

5.1 Gait

It is the most prominent and frequent (94–100% of patients) [32]. Magnetic gait or gait apraxia is symmetrical, otherwise it suggests other pathologies. Patients drag their feet, as if they are attached to the ground (hence, the term magnetic), the movement is clumsy and hasty, it is accompanied by small steps, frequent stumbling, falls and fear. Walking becomes unstable and slow. Difficulty standing and starting movement is characteristic, as well as erratic turns with many unstable steps, like a compass. Severity is variable and it is sometimes difficult to distinguish between parkinsonism and other dementias such as Lewy's. Unlike in Parkinson's disease, external triggers such as command lines and landmarks have little effect on improving gait [33].

5.2 Cognitive impairment

Cognitive decline in NPH is remarkably similar to that seen in other types of dementia. It includes infantile behaviors, mood fluctuations, amnesia, difficulty managing finances, taking medications, driving, and honoring commitments. Patient denial is common. Psychomotor speed is markedly declined, there is also evident attention and working memory impairment as well as diminished verbal fluency and dysexecutive syndrome [34, 35].

5.3 Incontinence

Urgency and frequency are the most common symptoms, initially without incontinence which appears progressively [36]. Amongst patients with NPH, 90.9% experience dribbling and 75% have incontinence [37]. They mostly are aware of this problem which causes great frustration. A patient with incontinence who is indifferent or unaware is unusual and should raise suspicion of another disorder. All this information is obtained from the patient's companion, usually his/her partner.

5.4 Other symptoms

There are other less frequent manifestations, which traditionally are not listed under NPH manifestations, therefore not commonly searched for, and not treated. An important example is hearing loss. Hypoacusis is present in a non-negligible number of patients with any type of chronic hydrocephalus. Thirty-four percent (34%) of patients who develop post-infectious hydrocephalus and a percentage that can range from 5 to 15% of patients with NPH have some degree of hypoacusis. Its mechanism is not well understood, but it could be secondary to hydrops of the endolymphatic and perilymphatic space, which is continuous with the SAS in the cochlear aqueduct in the posterior aspect of the petrous portion of the temporal bone [38]. In patients with NPH and hypoacusis, significant improvement has been observed after surgery [38]. Possibly if the problem is actively sought and specialized rehabilitation is offered, the outcome could improve significantly.

Approximately, 55% of NPH patients have bradykinesia, 84% have a snout reflex, 77% have an eyebrow reflex and 65% have some degree of paratonia [39]. In a study that compared patients with NPH, Parkinson's and healthy controls, the NPH group had decreased speed when lifting things and used more strength to grip compared to healthy controls. This correlates with involuntary motor dysfunction which resembles frequent neurologic deficit seen in Parkinson's.

5.5 Symptom assessment

The most important thing is to detect the components of the triad and to explore their characteristics in depth. As mentioned previously, other diseases should be ruled out [1].

When clinical suspicion is high, some important domains such as motor and cognitive performance should be evaluated profoundly. Different formal tests have been developed that allow objective and quantitative measures of compromise in all domains. We summarize tests that have been validated and recommended by clinical practice guidelines. Those tests are used in our NPH center of excellence (COE) at Fundación Santa Fe de Bogotá (FSFB), the only COE of its kind in Colombia currently accredited by Joint Commission International (JCI).

5.5.1 Gait

- Timed-up and go test: Time the patient takes to stand from the seat, walk 2 meters, turn 180°, walk back and sit down. This test has proved to be cost-effective, with high sensitivity, specificity, and positive predictive value (PPV) for favorable outcome after surgery [40].

- 10-meter test: Time the patient takes to walk 10 meters in a straight line starting in a standing position. Evaluation time is between meters 2 and 8 to minimize the effects of acceleration and deceleration [41].

- Tinetti: Test that integrates gait and balance with a maximum possible score of 30. The higher the score, the better motor performance. It has proven useful in evaluating response in patients undergoing CSF shunting surgery [42].

5.5.2 Cognitive decline

- Mini-Mental State Examination (MMSE): Widely known test that assesses various mental domains in 30 items including memory, praxis, literacy, and mathematical abstraction. Scores ≤23–25 suggest cognitive impairment and this score can be compared in the postoperative period with high specificity [40]. This type of test, although used for screening, is particularly useful in NPH patients with significant cognitive impairment.

- INECO frontal screening (IFS): Assesses executive processes that include motor response programming, inhibition of both motor and verbal responses, abstraction, verbal, and special working memory; also evaluates frontal dysfunction. The maximum score is 30 [43].

5.5.3 Incontinence

- ICIQ-SF (international consultation on incontinence questionnaire-short form): Validated questionnaire addressed to the patient and his/her companion evaluating frequency, severity, and impact of incontinence on the patient's quality of life. Score is 0–21, the higher the result, the more symptoms and the greater the negative impact on quality of life. There is a significant decrease in ICIQ-SF mean score after CSF shunting surgery [42].

5.5.4 Others

- FIM (functional independence measurement): Measures disability and level of dependence on another person in motor and cognitive domains that has shown high reliability [44].

- Zarit survey: Validated questionnaire to assess the degree of caregiver burden. It is done before and after surgery at each follow-up visit. The SINPHONI study demonstrated that the degree of caregiver burden decreased markedly after the patient's surgery and correlated with improvement in Hakim's triad symptoms [42].

6. Diagnostic tests

Despite the immense amount of research on NPH, to date, there is no gold standard test for its diagnosis. Therefore, diagnosing NPH is a challenging task that results from gathering information on clinical manifestations, radiographic findings, clinical response after tap test, and recently, certain CSF biomarkers.

6.1 Imaging

After the aforementioned tests, a CNS image is necessary. Of choice, brain magnetic resonance imaging (MRI) or computed tomography (CT) if MRI is contraindicated or not feasible; MRI is preferred due to its higher resolution and definition of parenchymal structures whether healthy or pathological.

According to the Japanese Guidelines for management of Idiopathic Normal Pressure Hydrocephalus, the imaging findings that suggest NPH are: ventriculomegaly not attributable to other pathology, disproportionately enlarged subarachnoid space hydrocephalus (DESH), acute callosal angle and posterior narrowing of the cingulate sulcus seen on sagittal plane MRI [32].

The main finding is ventriculomegaly assessed with the Evans' Index (EI), which is calculated by measuring the maximum width of both frontal horns of the lateral ventricles and dividing it by the maximum intracranial width in the same slice of axial plane; a normal value is <0.3 (**Figure 1a**). A newly proposed z-Evans' Index has shown an increased diagnostic value, it is measured in the coronal plane and is a ratio of the maximum frontal horn height and the vertical diameter of the skull at the midline, the normal value is <0.42 (**Figure 1b**) [45, 46]. The distribution of subarachnoid spaces changes with NPH as well as neurodegenerative diseases such as AD and Parkinson's disease. A characteristic pattern that aids in differentiating NPH from other pathologies is DESH, its imaging features are narrowed subarachnoid spaces of the midline and high convexity with dilation of the Sylvian fissures, associated with ventriculomegaly (**Figure 1c**). In contrast,

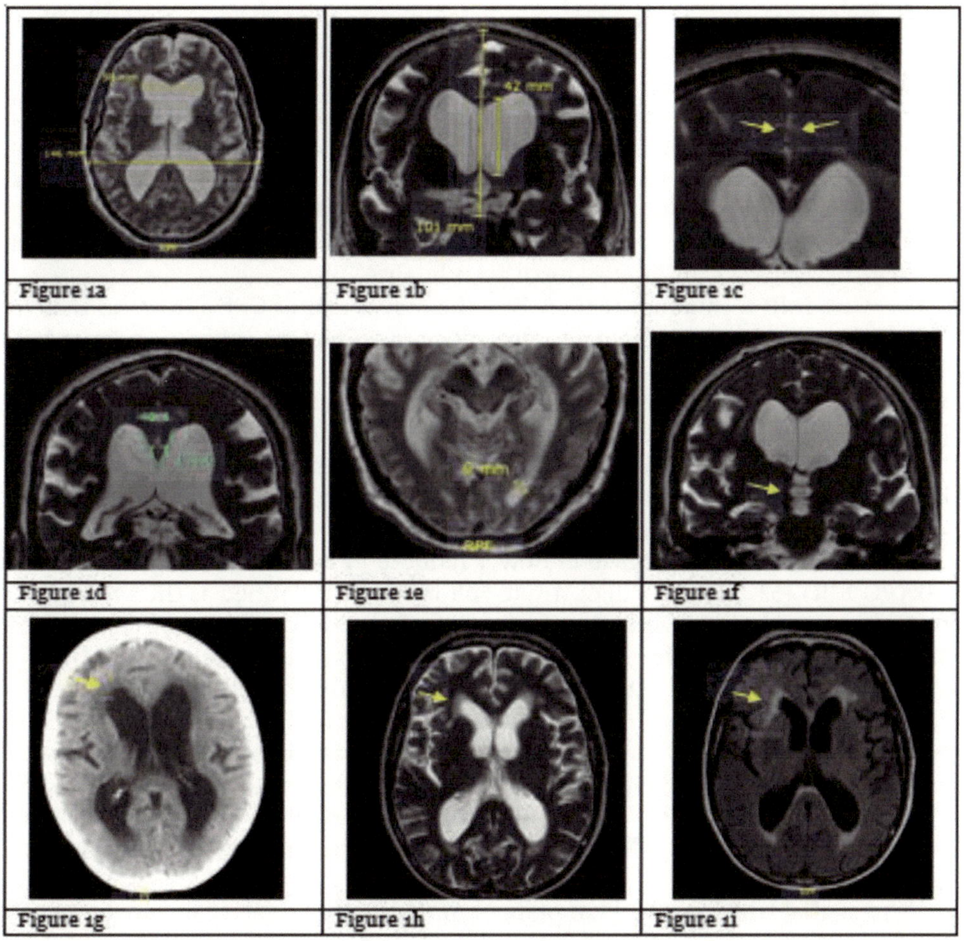

Figure 1.
NPH imaging findings. a. T2WI axial MRI. Evans' index b. T2WI coronal MRI. Z-Evans' index c. T2WI coronal MRI. DESH d. T2WI coronal MRI. Callosal angle e. T2WI axial MRI. Temporal horns f. T2W1 coronal MRI. Ventricle ballooning g. axial CT. Transependymal CSF flow h. T2WI axial MRI. Transependymal CSF flow i. T2WI Flair. Transependymal CSF flow. Images obtained from FSFB NPH COE database.

generalized widened sulci with ventriculomegaly is seen on neurodegenerative cerebral atrophy [47]. Due to morphological changes attributed to DESH, vertical distortion of the lateral ventricles is more commonly associated with NPH and is evidenced in the CA, which is the angle formed by the right and left parts of the corpus callosum and is measured in a plane perpendicular to the anterior commissure-posterior commissure plane (AC-PC) through the posterior commissure; patients with NPH have an acute angle (<90°) whereas healthy patients or patients suffering from other diseases have an angle >90° (**Figure 1d**). Proper alignment of the true perpendicular plane must be achieved to avoid over or underestimation of the CA [48]. A systematic review conducted by Park et al. analyzed the diagnostic performance of the EI and CA and demonstrated that the CA yields a higher diagnostic value than EI (CA: 91% sensitivity and 93% specificity versus EI: 96% sensitivity and 83% specificity) [49].

A less common method is to measure the thickness of both temporal horns of the lateral ventricles in an axial section and if this value is >2 mm it is highly suggestive of ventriculomegaly (**Figure 1e**). Lateral ventricle ballooning and/or slit-shaped third ventricle also suggest ventriculomegaly (**Figure 1f**). In cases of third and fourth ventricular dilation, it is important to rule out triventricular or tetraventricular hydrocephalus due to mechanical obstruction of another cause [50]. Other signs are low periventricular density on CT or hyperintensity on T2WI or FLAIR MRI sequences immediately adjacent to the ventricular wall, which is highly suggestive of transependymal flow of CSF (indirect sign of hydrocephalus) (**Figure 1g–i**). More peripheral white matter lesions, such as those in corona radiata, suggest ischemic changes [50]. It is important to differentiate ventriculomegaly in NPH from compensatory *ex vacuo* ventriculomegaly secondary to atrophy, typical of other diseases. Focal atrophy indicates other types of dementia, especially if it is asymmetric, as in FTD or hippocampal atrophy in AD.

There are MRI techniques that analyze the movement of CSF and water that offer further insight about the pathophysiology of NPH. One is phase contrast MRI (PC-MRI) which is an invasive technique that measures flow of ejected CSF in specific anatomic regions, especially the cerebral aqueduct. It is based on the pulsatile relationship of CSF flow with heartbeat and respiratory rate and known relaxation times of CSF in T1WI and T2WI sequences with respect to static tissue. Limitations include that it is invasive, gathers information from many cardiac cycles, and evaluates only certain regions [51]. Another reported issue in NPH is increased CSF pulsatility. Aqueduct flow is reduced or even absent in phase contrast MRI (PC-MR) [52–54]. Aqueduct stroke volume (ASV), defined as the average of flow volume through the aqueduct during diastole and systole, may be an indirect parameter of CSF pulsatility. Various studies have shown increased ASV in NPH patients compared to healthy controls. Luetmer et al. demonstrated that ASV elevation assists in diagnosis and in differentiating NPH from other dementias. ASV greater or equal to 42 L has been applied to identify patients who might benefit from shunting procedure [55]. Scollato et al. reported that patients with higher ASV may benefit from shunting [56]. Aqueduct pulsatility reflects capillary expansion, mainly influenced by the pulsatile dampening of arteries in the Windkessel *effect* [57, 58].

Time-spatial labeling inversion pulse or *SPLIT*, a non-invasive technique that synchronizes CSF flow with the arterial pulse. SPLIT evaluates both linear and turbulent movement of CSF between two compartments in any anatomic region of the CNS for periods of up to 5 seconds (22). The observed flow patterns are not the same as those classically described, even in brains without hydrocephalus. In fact, retrograde aqueduct flow generates sustained pressure gradients that favor compressive stress and shearing force on the ependyma. The causal relationship between ASV and ventricular volume has been confirmed [59]. Further improvements in this technology will probably offer valuable information on the pathophysiology of NPH in the short-coming future.

Arterial spin-labeling or ASL-MR perfusion is a non-invasive technique without the need of intravenous contrast administration. Instead, ASL-MR perfusion technique is based on the principle of magnetically labeled water molecules in blood, hence, water in blood entering the CNS is used as an endogenous tracer. Unlabeled images are subtracted from labeled images obtaining a regional blood flow map. Some authors have used ASL-MR to study changes in cerebral flow in patients with NPH and have found a positive correlation with clinical changes before and after surgery [8]. Diffusion tensor imaging (DTI) has gained popularity for the diagnosis of NPH, following dilation of the lateral ventricles comes abnormalities and compression of the surrounding periventricular white matter. The research on the usefulness of DTI in NPH has been focused on its diagnostic value and ability to differentiate NPH from other neurodegenerative diseases, the areas of interest are the corpus callosum, internal capsule, hippocampus and the corticospinal tracts to which gait disturbance could be attributed. Fractional anisotropy (FA) which is a measure of the direction of diffusion, is reported in absolute values ranging from 0 to 1 where 0 means free flow of molecules and 1 means restricted linear flow. Increased FA signal is reported as a prominent characteristic in NPH compared to other neurodegenerative diseases like AD or Parkinson's disease where FA is decreased. However, recent research reports decreased FA of the corticospinal tracts in NPH [60].

Single photon emission computed tomography (SPECT) can be used for analysis of cerebral blood flow (CBF) when studying NPH, the characteristic CBF pattern observed in NPH is known as convexity apparent hyperperfusion (CAPPAH) which consists in increased CBF in midline and convexity with decreased perisylvian perfusion congruent with DESH morphology [32, 46]. Glucose metabolism assessed with Positron Emission Tomography is useful for distinguishing NPH from other pathologies when decreased basal ganglia metabolism is present. MRI flowmetry and spectroscopy have not been shown superior to the aforementioned imaging methods, therefore cannot be recommended [32]. Other nuclear medicine techniques such as dopamine transporter scintigraphy, fluorodeoxyglucose-PET (FDG-PET) and amyloid-PET have also been studied in NPH offering possible diagnostic aids.

Given the interobserver variation in the findings on brain imaging suggesting NPH, an objective standardized guideline or scale would be warranted. A radiologic scale (iNPH Radscale) was proposed in 2017 by Kockum et al., evaluating the correlation of NPH symptoms and seven radiological findings in CT: EI, CA, DESH (Sylvian fissure dilation and narrowing of parafalcine or high convexity sulci), temporal horn enlargement, focal enlargement of sulci and periventricular hypodensities, with a score ranging from 0 to 12. The study results showed that a higher score was related with a higher symptom burden, however the cohorts studied were only from a Swedish town [61]. In 2020 the diagnostic performance of the iNPH Radscale was evaluated by its author, the follow-up results showed that a cutoff value of 4 had high diagnostic value with a sensitivity of 100% and specificity of 96%, which could be a useful diagnostic tool in the future when further studies are conducted with regards of its diagnostic values in other populations [62].

6.2 Tap test

One of S. Hakim's most important lesson is that lumbar puncture (LP) is a technically simple procedure, with roughly null complication rate and cost-effective that provides valuable information on neurophysiology. Tap test simulates a drainage device improving NPH symptoms. However, it has a low sensitivity and there is no consensus on the parameters that should be used nor the volume to be extracted. In our NPH COE at FSFB, all suspicious cases of NPH undergo a modified tap test that consists in a conventional LP: the patient is in a lateral decubitus position and an 18-gauge (18.0 Ga) spinal needle is used. The amount of CSF obtained is determined by reaching a closing pressure of 0 cm H_2O regardless of the volume, thus there is no fixed volume to extract. We recently carried out a descriptive cross-sectional study in which 92 patients with a mean age of 79.4 years were included. The diagnosis was confirmed in 73.9% of cases (validity comparable to that reported in the literature). The mean opening pressure was 14.4 cm H2O and the mean volume extracted was 43.4 mL [63]. These results warrant further trials to determine if this modified tap test could eventually become the gold standard. If opening pressure is normal and there is objective improvement in symptoms and formal tests 24 hours after LP, the patient is considered candidate for CSF shunting surgery.

6.3 Lab tests

Although most patients have one or more symptoms of the clinical triad, as well as typical findings in imaging, and procedures that can lean towards the diagnosis of NPH, no gold standard has been established. For this reason, since 1990 some authors have studied the role of biomarkers in serum and CSF that had not been frequently used before despite their proved diagnostic value in other neurodegenerative disorders, especially AD. Results using serum biomarkers were not promising, and the authors concluded that as long as the BBB is intact, as it would be expected in NPH, serum levels of those markers would not have diagnostic utility [64–66].

To date, there is no consensus on the biochemical profile of NPH and the differences with the profile of some neurodegenerative disorders, but interesting data has been obtained on some biomarkers:

Amongst neurotransmitters, activity of AChE in CSF has been positively correlated with MMSE scores in patients with NPH and AD, which is reduced in these diseases [67–69]. Other neurotransmitters have not been proven to be useful due to their ubiquity in the CNS and poor correlation.

Somatostatin (SOM) levels have been described as lower in NPH patients compared to controls and these levels have increased after CSF shunting surgery. Increase in SOM levels correlates positively with cognitive performance (memory and visuo-motor) after surgery, but this does not persist over time [70, 71]. Neuropeptide Y (NPY) levels are also reduced in NPH compared to controls and increase after surgery, although they are also low in patients with AD, so this finding is not specific of NPH [71, 72]. Vasoactive intestinal peptide (VIP) is markedly elevated in patients with SAD or Binswanger's disease and in general in vascular dementias. VIP levels are higher in patients with NPH and cerebrovascular comorbidity compared to a NPH group without any comorbidities [73].

Tumor necrosis factor alpha (TNF-alpha) levels may be increased up to 45-fold above reference range in NPH patients and interestingly, TNF-alpha levels normalize after surgery, suggesting a pro-inflammatory component not yet studied in NPH. Its half-life is short and therefore researchers highlight it is not related to stagnant CSF or siphoning effect after implantation of the shunting device. It could be a specific marker with a high PPV [74]. Vascular endothelial growth factor (VEGF) is a proangiogenic cytokine that serves as a marker for chronic hypoxia. Its levels are elevated in patients with NPH compared to controls; the higher VEGF levels, less response to surgery and worse clinical outcome [75, 76].

Myelin basic protein (MBP) is a reliable marker of demyelination, and hydrocephalus is known to cause periventricular demyelination by mechanical stretch, hence, the degree of ventriculomegaly has been positively correlated with MBP levels. Brain atrophy, though, does not raise MBP levels [77, 78]. These levels decrease after surgery, which supports the theory that ventriculomegaly in NPH does not produce atrophy but cerebral pseudoatrophy, studied by some authors including Fernando Hakim et al. and correlates with clinical improvement after surgery [79–83].

In a study of experimentally-induced hydrocephalus in rats, postoperative MBP levels were lower in models operated 1 week after induced hydrocephalus compared to models operated 4 weeks later [84]. This supports the idea that the earlier hydrocephalus is diagnosed and treated, the better the outcome for NPH patients.

Beta-amyloid is a peptide synthesized from amyloid precursor protein (APP) whose biologic functions include enzymatic co-factor, cholesterol transport, and pro-inflammatory activity. The diagnostic role of beta-amyloid deposits in AD has been largely studied. Especially the ratio of its isoforms (AB42 / AB40) which has a higher diagnostic value than AB42 alone. The levels of AB42 and AB40 isoforms are low in NPH compared to AD [84, 85]. They could be helpful in distinguishing between NPH and AD. Nearly half of NPH patients show certain degree of amyloid deposition in brain biopsy specimens while only 10% show concomitant amyloid and tau pathology [86]. Tau is a protein of neural tissue microtubules and serves as a marker of neuronal degeneration in other types of dementia like AD, Lewy's, corticobasal degeneration, and prion disease (Creutzfeldt-Jakob). The levels of total tau (t-tau) and phosphorylated tau (p-tau) have been positively correlated with the severity of dementia in AD, as well as with cognitive decline and urinary incontinence in NPH, although they are lower in NPH compared to AD [84, 87]. Neurogranin (NRGN) has also been studied in AD patients and has been associated with amyloid plaques [88, 89].

Neural growth factor (NGF) mRNA levels are elevated in the basal nuclei of experimentally-induced hydrocephalus models, this is why they could act as markers of early neuronal injury that stimulates glial recruitment [90]. Neurofilament light chain (NFL) is increased in NPH, AD, and other dementias, but there is no correlation with clinical manifestations, severity, or response to surgery. A Swedish study reported 100% PPV for positive outcome after surgery, with 17% sensitivity and 100% specificity [91, 92]. It could be a marker of ongoing axonal damage.

Sulfatide is a glycosphingolipid component of myelin. Its levels are higher in NPH and cerebrovascular comorbidity, but no correlation with postoperative outcome has been reported. According to one study, sulfatide differentiates NPH from SAD with 74% sensitivity and 94% specificity [93]. Glial fibrillary acid protein (GFAP) and S-100 are markers used in tumor immunohistochemistry that have not shown conclusive data in NPH.

Recently, the first longitudinal comparison of CSF biomarkers in NPH patients and AD was conducted by a group in Finland. Furthermore, they compared the gradients of biomarker concentrations lumbar and ventricular CSF which are discordant. All markers increased notably by 140–810% in lumbar CSF, except beta amyloid that had an erratic behavior. All studied biomarkers (tau, NFL, NRGN and beta amyloid) correlated highly between lumbar and ventricular samples but were systematically lower in ventricular samples [94]. Longitudinal follow-up shoed that after initial postoperative increase, tau and NRGN levels are stable in NPH regardless of brain biopsy amyloid pathology. NFL normalized after surgery to pre-shunting levels. Amyloid is the less affected by shunting and may be the best predictor of concomitant AD risk in NPH patients [94]. Tau levels (both p-tau and t-tau) have shown a steady increase of 2% per year in AD [95]. In cases of TBI, t-tau levels increase but normalize to baseline levels by day 20–43 post trauma [96, 97]. This is different, however, for patients who suffer chronic TBI like boxers, for example [97]. Interestingly, NFL levels after surgery correlated with tau and NRGN, raising the question if their increase is related to disease process or the shunting procedure itself. Future larger randomized trials are warranted to elucidate the role of CSF biomarkers in NPH progression, surgical prognosis, and risk of concomitant neurodegenerative disorders.

7. Brief history of CSF shunting devices

The main principle is shunting CSF to another sterile cavity with constant flow and/or where fluid can be reabsorbed to the systemic circulation. The quest for an effective system is not recent. Le Cat performed the first documented ventricular puncture in 1744 [98, 99]. Throughout the XIX and XX centuries, different devices were designed as well as different shunting techniques that included ventriculo-subarachnoid-subgaleal, lumbo-peritoneal, ventriculo-peritoneal, venous, pleural, and uretheral. All failed in the short-term because implant materials that included glass, rubber and guttapercha were not adequate and aseptic techniques were extremely poor with high rates of infection [99]. There were also cases of sudden death before the advent of appropriate imaging techniques which fortunately decreased after Dandy introduced the pneumo-ventriculography in 1918. Pneumo-ventriculography remained the gold standard until 1980 when Humphrey introduced the computed tomography [99]. During the first half of the XX century different techniques were proposed such as choroid plexus cauterization and draining systems to intra and extracranial veins. Infection and material rejection were the main causes of failure. Developing a biocompatible unidirectional system was paramount.

Torkildsen described the ventriculo-cysternostomy in 1938 in a case of spontaneous cure of hydrocephalus after an accidental rupture of the fourth ventricle during surgery. This became a popular method to treat obstructive hydrocephalus until the early 1970's [100]. Vannevar Bush engineering professor at Massachussets Institute of Technology (MIT) and Donald Matson, surgeon at Harvard Children's Hospital were possibly the first to develop a magnetically-operated valve by 1950. Although there is no certainty on the exact date of implantation, around 18 devices were implanted by 1957 but this project was soon abandoned because results were not promising [99]. Ommaya invented the subcutaneous reservoir in 1963, his device is still today the method of choice for obstructive hydrocephalus in children with minor modifications to the original design [101].

7.1 First effective devices

The long-sought biocompatible material is silicone, an inorganic polymer derived from polysiloxane (series of oxygen and silicon atoms) whose properties include inert, malleable, resistant to high temperatures and stretching. Silicone has its origins in World War II, like many other inventions, because materials that resisted high temperatures, provided electrical insulation, and resisted mechanical stress for aircraft construction were urgently needed. In 1946, a silicone tube was implanted to repair a bile duct. In 1956, it was used as a CSF drainage device by Holter and Pudenz and thus became the ideal material for different valve designs [99]. However, these devices failed because, due the mechanism of the opening slit when CSF pressure increased, they were imprecise.

In 1964, S. Hakim introduced his first valve that consisted of two twin valved systems made of stainless steel and synthetic sapphire manufactured by himself in his home lab. It was one of the first precise models that controlled pressure [102]. By 1965, S. Hakim had published his observations in the NEJM and NPH or Hakim's syndrome was recognized as a separate entity by the scientific community. A second generation of devices intended to solve the problem of overdrainage in the upright position. Kuffer and Strub designed a piston-based system in 1969, but it never became popular and the same happened with some successive designs. S. Hakim introduced a valve that could be operated magnetically and percutaneously in 1973 [99]. In that same year, he introduced a self-regulating device, nonetheless, the first patented self-regulating valve was that of Sainte-Rose in 1984 (Cordis Orbis-Sigma). Portnoy later designed his *anti-siphon* device patented by Schulte in 1973 (Heyer-Schulte ASD). This is a flapping membrane mechanism that closes progressively when subjected to the weight of a hydrostatic column in the distal catheter. These mechanisms failed because they were highly susceptible to external tissue pressure. In 1975, S. Hakim patented the first *anti-gravitational* device [99].

At the time, S. Hakim's oldest son, Carlos, was already a mechanical engineer starting his post-doctoral fellowship at MIT in biomedical engineering. Carlos studied hydrocephalus in animal models. In his thesis he questioned some aspects of his father's original theory including the concept of brain elasticity. Carlos demonstrated that if the brain were elastic, dilated ventricles would return to normal dimensions after a shunting device was implanted. This does not happen in all cases and it depends on how much time ventricles have been subjected to the *hydraulic press effect*. He introduced the concept of brain *plasticity*. Carlos Hakim also demonstrated that the thin walls of the venous system in an adult human are easily compressible when subjected to high external pressure in the upright position. Thence, demonstrated that the anti-siphon effect was not true. Instead, he associated with Swiss watchmakers who were pioneers in micromechanics and developed the first *programmable* valve with 18 pressure positions ranging from 30 to 200 mm H_2O (Medos-Hakim). This device was first implanted in Colombia by S. Hakim and was approved for commercialization in 1989. It soon demonstrated superiority compared with all other available designs and entered Europe in 1990 [99]. Today, this design is distributed by Johnson & Johnson as the Codman Hakim™ programmable valve which has proved superiority systematically.

By 1999, at least 127 models of different mechanisms were available. Most of them rudimentary unidirectional systems based on pressure gradient with ball and cone (13 models), diaphragm (>35) and slit (>50) [99]. To date, >60 models were never evaluated in lab and 40 were only in tested in one or two specimens, hence, their value is only anecdotal. Compared with other high-tech devices in the biomedical field, like pacemakers, most valves are imprecise, unsafe, obsolete, and cheap. By 2000, the mean cost per device was $600 USD. Assuming a mean device lifespan of 10 years, this equates to 17-dollar cents/day. The total cost of sold devices in the USA by 1995 was 20.8 million dollars which was the equivalent of 8-dollar cents/per capita [99, 103].

8. Surgical treatment

8.1 Surgery

To date, the only effective treatment of NPH is a CSF shunting procedure that involves implantation of a draining system that diverts excess intracranial CSF into a sterile cavity where CSF returns to or is reabsorbed into the systemic circulation.

The system should ideally be a programmable valved device. Shunting techniques include VAS, VPS, LPS, and rarely ventriculo-pleural shunt (VPlS). VAS and VPS are the most commonly used in clinical practice. S. Hakim originally described VAS arguing that it was physiological and of choice at our NPH COE. LPS is very common amongst Japanese surgeons, however, it is technically complex and can have a higher rate of perioperative complications. DVPl is seldom used due to a high complication rate and lower reproducibility. It is consensus that endoscopic third ventriculostomy (ETV) is not effective for treating NPH.

8.1.1 VAS

It is considered the technique of choice. Between 1970 and 1980 it was hardly criticized because it was an expensive technique that required vascular dissection, hence, with multiple complications such as vascular rupture, embolism, and infection. However, with the advent of Seldinger-type techniques guided by ultrasound (US) and constant electrocardiographic monitoring, which do not require vascular dissection, incidence of complications in VAS does not exceed that of VPS [104, 105].

At our NPH COE, VAS is performed in all patients unless contraindicated. The technique consists of puncturing the internal jugular vein (IJV) using a US-guided Seldinger technique and continuous EKG monitoring. A 7-Fr peel-away disposable sheath catheter is used. This allows easy, safe, fast, and reproducible insertion and positioning of the distal catheter in the cavo-atrial junction which is the correct position (guided by fluoroscopy) [105]. Trained surgeons can perform the whole procedure within 30 minutes, with a lower perioperative complication rate than other techniques. Hung et al. reported that patients with NPH undergoing VAS are less likely to develop obstruction and/or require device revision compared to the group undergoing VPS [105]. Serious thromboembolic complications associated with VAS such as in situ thrombus formation, intracardiac thrombi, pulmonary artery thromboembolism and pulmonary hypertension, which have a high morbidity and mortality rate, are uncommon, with an estimated prevalence <1%. Adjuvant therapy with direct anticoagulant agents like rivaroxaban has been proposed, but currently, there is no robust evidence to support such recommendation as a preventive strategy [105–107].

8.1.2 VPS

Became the most popular technique since 1970 because of the vascular complications encountered in VAS. However, there is a non-negligible percentage of patients in whom it is advisable not to use a distal peritoneal catheter due to inflammatory or infectious diseases in the abdominal cavity, as well as a slightly higher incidence of distal obstruction. Patients undergoing VPS, have an estimated incidence of device-related complications in the peritoneal cavity ranging from 5 to 47%, depending on the series. Complications include device infection, pseudo-cyst, adhesion, and malposition (scrotum, bladder, small intestine, and hernia) [108].

8.1.3 LPS

Murtagh was the first to introduce a lumbo-peritoneal catheter through a Touhy needle [83]. Rarely used amongst Western surgeons, but extremely popular amongst Japanese neurosurgeons. Described as a less invasive and effective alternative in high-risk patients in whom the right atrium and peritoneal cavity cannot be used. Significant improvement has been reported in the components of Hakim's triad, however, the complication rate is 20% [28, 109, 110]. This technique is not routinely recommended unless the surgical staff is highly experienced.

8.1.4 VPlS

An alternative when VAS or VPS are contraindicated, however, the complication rate is high including hydrothorax, pneumonia, and pleural effusion. It is not considered an efficient long-term option [111].

8.1.5 ETV

Described in 1990 as an alternative approach for cases of obstructive hydrocephalus and some selected cases of communicating hydrocephalus, has recently gained attention as an alternative approach that saves device implantation. However, conducted trials have small numbers of patients and lack randomization, therefore its effectiveness and generalizability in NPH is still limited [112, 113].

The thorough description of each surgical technique is beyond the scope of this chapter.

8.2 Prognosis after shunting procedure

Despite variability in evaluation methods, gait disturbance has systematically showed the highest rate of improvement (60–77%). Cognitive decline improves in 60–70% of cases and urinary incontinence improves in 52% of cases [114–118]. Surely, these rates vary according to different diagnostic criteria, evaluation methods and improvement thresholds.

Short-term outcomes are mainly affected by perioperative complications and by severity and duration of disease before treatment. Long-term outcomes strictly depend on other neurologic comorbidities, hence the importance of diagnosis these disorders using biomarkers before and after surgery to help patients and their families conveying their expectations. Frailty and/or comorbidity indices can be helpful for perioperative outcome evaluation.

8.3 Shunt procedure cost-effectiveness

Based on SINPHONI and SINPHONI-2 results, incremental cost-effectiveness ratio 1 year after shunt surgery was 29934–40742 USD/quality-adjusted life year (QALY) for VPS and 58346–80392 USD/QALY for LPS. Additionally, the sum of surgical cost and nursing cost for NPH is reduced to 18 months after VPS and 21 months after LPS, compared with untreated NPH patients [119].

Some authors studied the economic effect of NPH treatments using the Markov model based on epidemiologic data from Sweden. These authors reported that an additional lifetime of 2.2 years and 1.7 QALY was gained with treatment, with an additional cost of 13,000 GBP [120].

8.4 Non-surgical treatment

To date, there are no FDA-approved pharmacological therapies for NPH. Clinical trials suggest that carbonic anhydrase inhibitors (CAIs) such as acetazolamide can reduce periventricular white matter hyperintensities and thence improve NPH symptoms [121]. However, these studies have flawed designs which confound their results and render their conclusions temporarily invalid. Prospective, double-blinded, and placebo-controlled trials are warranted.

Future improvements in technology within the pharmaceutical industry may offer novel supplementary agents that tackle NPH pathogenesis. These drugs could normalize CSF hydrodynamics by tackling CSF production, pulsatility and Rout. They could also restore cerebral blood perfusion and parenchymal compliance as well as promote brain waste products providing neuroprotection and reducing neuroinflammation.

9. Complications after CSF shunting surgery

Main complications include infection, catheter malposition (proximal and distal), and hydraulic device-associated complications. Infections are responsible for 10–15% of device revisions, but their impact on morbidity and mortality is high. The cost of managing a patient with an infected device is approximately $30,000 USD [103]. Most infections are secondary to intraoperative contamination of the implant [99]. Trained surgical staffs that meet strict aseptic techniques and are able to reduce operating time have infection rates <1%. In 3 independent meta-analyses, the use of prophylactic systemic antibiotic decreased infection rate to 5–6% compared to 10–12% in controls without prophylaxis [122–124]. At our NPH COE, intravenous Vancomycin is infused during a 60-minute period before incision.

Catheter malposition, both proximal and distal, is the most common cause of shunt system failure, but it is an unpopular topic in the literature [125]. Probably due to ambiguity in the definition of perioperative complication and the methods used to analyze them. At some point in the post-operative follow-up period, approximately 17% of patients develop complications associated with overdrainage such as hygromas and subdural hematomas, postural headache, slit ventricular syndrome and, occasionally, bone table deformities. Other less frequent ones include proximal occlusion, sequestered ventricle, upright ventricular hyperemia, intraparenchymal or intraventricular hemorrhage and/or partial sometimes irreversible loss of cerebral compliance [126]. Postural headache along with hygromas and laminar subdural hematomas, can be managed with pressure adjustment of the device. According to a recent metanalysis, the need for additional surgery was 9–16% of patients operated with an adjustable device and 26–38% in patients with a fixed-pressure device [127]. These are the main reasons to use programmable devices because percutaneous pressure adjustment saves additional surgical interventions for complications that can be resolved non-invasively. When the hygroma or subdural hematoma is large, it produces intractable headache and/or progressive neurologic deficit, surgery to drain the space-occupying lesion, device revision and proximal catheter relocation is indicated.

Novel shunt catheters manufactured with advanced biomaterials that avoid cell and bacterial adhesion as well as *smart* devices with auto-regulating monitors are underway which promise better treatment outcomes [128].

10. NPH COE protocol

At FSFB, we designed a NPH protocol based on the best available evidence and standards. Our protocol has been certified and accredited by Joint Commission International (JCI). The protocol consists of 5 phases with specific goals to diagnose and treat NPH in an optimal way. The main purpose of this protocol is helping the patient reintegrate to his/her daily activities and community (**Figure 2**).

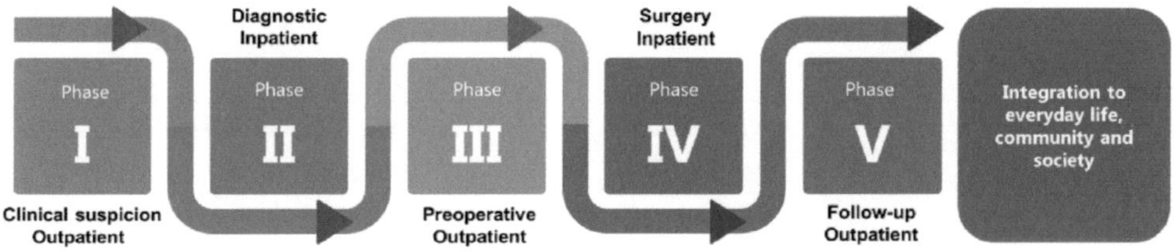

Figure 2.
FSFB NPH COE protocol.

Gait impairment (praxia/symmetric magnetism) or cognitive decline or urinary urgency/incontinence
Radiographic signs suggestive of NPH
Absence of other neurologic disorder that could explain current symptoms

Table 1.
Inclusion criteria. Patients should have at least one of the following (≥ 1).

10.1 Phases

10.1.1 Phase 1 (clinical suspicion)

Patients either consult or are referred because have signs and symptoms suggestive of NPH. The objective of this phase is to identify suspicious cases. Patients undergo a complete neurologic examination and interview by one neurosurgeon of our team. The physician orders a brain MRI or CT scan whenever MRI is contraindicated. Suspicious cases are those who meet the inclusion criteria listed in **Table 1**.

10.1.2 Phase 2 (diagnostic tests)

Patients who met the inclusion criteria and consent are admitted for a 2-day inpatient analysis. The patient undergoes thorough evaluation by neuropsychology, speech and language therapy, rehabilitation medicine and geriatrics specialist using all diagnostic tests that have been mentioned before. In all patients >64 years, an elder adult frailty test is performed. Depending on each individual case, social work and nutrition specialist may also be involved. Then we perform the modified tap test we described above by taking the system to a closing pressure of 0 cm H2O. All tests are repeated 24 hours after tap test. Tests performed before and after tap test are listed in **Table 2**. Thereafter, the patient is discharged and followed during 7 days via phone call to inquire on symptom improvement.

After this week of follow-up, a multidisciplinary team consisting of all neurosurgeons, psychology, speech and language therapy, rehabilitation medicine and geriatrics meets to analyze objective changes in all tests before and after tap test. Diagnosis is confirmed when patients objectively improve in at least 1 of the triad components. Diagnostic criteria are listed in **Table 3**. Once the diagnosis of NPH is confirmed, the team decides if the patient benefits or not of a CSF shunting surgery. When NPH is discarded, patients are referred to the required specialty.

Patient companions undergo the Zarit caregiver's burden test before and after tap test to evaluate the degree of burden/fatigue.

10.1.3 Phase 3 (preoperative assessment)

If patient and family agree with the team's decision, the patient undergoes regular preoperative tests, anesthetic evaluation, and an institutional risk mitigation

Area	Test
Physical rehabilitation	FIM
Physical therapy	Tinetti assessment tool
Occupational therapy	• MMSE • Pegboard test
Neurosurgery	Gait: • Timed-up and go. • 10-meter test Incontinence: • ICIQ-SF Burden: Zarit
Neuropsychology	• IFS • Stroop's color test • CERAD/ADAS-COG • Rey-Osterrith's complex figure

Table 2.
Tests performed before and after tap test.

Time ratio between any gait test before (t1) and after (t2) tap test greater than 1: $t1/t2 > 1$
Subjective improvement in gait symptoms after tap test according to specialist
Objective improvement after tap test in at least one neuropsychological test
Objective improvement in FIM scale
Subjective improvement after tap test according to patient companions
Objective improvement in ICIQ-SF score after tap test
Normal CSF pressure during tap test
At least one of the following: EI > 0.3, periventricular transependimary flow and/or convexity tightness

Table 3.
Diagnostic criteria at FSFB NPH COE.

form. Patients are then scheduled for CSF shunting. Patients and family receive rigorous education on the procedure and postoperative care.

10.1.4 Phase 4 (surgery)

Patients are admitted for VAS or VPS when VAS is contraindicated.
All cases include:

- Prophylactic IV vancomycin one hour before incision. If allergy to vancomycin is present, clindamycin is used.
- All patients are operated using a programmable device.
- The technique of choice is a Seldinger-type US-guided VAS using a peel-away 7-Fr disposable catheter. Correct position is confirmed using fluoroscopy.
- Device opening pressure is set according to intraventricular opening pressure (IVOP). The device is programmed 10 cm H2O below this value.

After surgery, patients stay for 48 hours as inpatient for close postoperative follow-up. During these 48 hours, physical rehabilitation specialist designs an individual rehabilitation plan. Again, education on device care, hygiene and urgent signs that may require visit to the emergency room are provided.

Patients are discharged on the third postoperative day. A control CT scan is ordered, and patients are instructed to visit their surgeon on the tenth postoperative day at the outpatient clinic.

10.1.5 Phase 5 (follow-up)

Patients are followed on months 1, 3, 6 and 12 after surgery. The main objective is to evaluate shunting device functionality and symptom improvement. Psychology tests are performed on months 6 and 12. Rehabilitation plans are adjusted according to every individual's needs.

11. Conclusions

Despite the huge amount of research since its original description, NPH remains an underdiagnosed disorder due to lack of clinical suspicion amongst the medical community. Therefore, the main concern is raising interest and suspicion amongst all medical professionals.

Recent advances in diagnostic imaging and lab biomarkers have given interesting insight into the pathophysiology of NPH that will probably be fundamental in the future. The best proven method for diagnosing and treating patients with NPH is following a standardized multidisciplinary protocol.

Conflict of interest

None of the authors have any conflict of interest to disclose.

Abbreviations

AChE	Acetyl cholinesterase
AC-PC	Anterior commissure-posterior commissure plane
AD	Alzheimer's dementia
APP	Amyloid precursor protein
AQP4	Aquaporin 4 channels
ASL-MR	Arterial spin-labeling magnetic resonance
ASV	Aqueduct stroke volume
BBB	Blood–brain barrier
BuChE	Butyryl cholinesterase
CA	Callosal angle
CAI	Carbonic anhydrase inhibitor
CAPPAH	Convexity apparent hyperperfusion
CBF	Cerebral blood flow
CSF	Cerebrospinal fluid
CSFBB	Blood-cerebrospinal fluid brain barrier
CNS	Central nervous system
COE	Center of excellence
CPP	Cerebral perfusion pressure
CT	Computed tomography

CVP	Central venous pressure
DESH	Disproportionately enlarged subarachnoid space hydrocephalus
DTI	Diffusion tensor imaging
DSIP	Delta sleep-inducing protein
EI	Evans' index
ETV	Endoscopic third ventriculostomy
FA	Fractional anisotropy
FDA	Food and Drug administration
FDG	Fluorodeoxyglucose
FIM	Functional independence measurement
FLAIR	Fluid attenuated inversion recovery
FSFB	Fundación Santa Fe de Bogotá
FTD	Frontotemporal dementia
GABA	Gamma-aminobutyric acid
GBP	Great British pound
GFAP	Glial fibrillar acid protein
5-HIIA	5-Hydroxyindoleacetic acid
HVA	Homovanillic acid
ICH	Intracranial hypertension
ICP	Intracranial pressure
ICP_{SA}	Subarachnoid space intracranial pressure
ICP_{SS}	Sagittal sinus intracranial pressure
ICIQ-SF	International consultation on incontinence questionnaire – short form
IFS	INECO frontal screening
IVOP	Intraventricular opening pressure
JCI	Joint Commission International
LP	Lumbar puncture
LPS	Lumbo-peritoneal shunt
MAP	Mean arterial pressure
MBP	Myelin basic protein
MHPG	3-methoxy-4-hydroxyphenylglycol
MIT	Massachussets Institute of Technology
MMSE	Mini-mental state examination
MRI	Magnetic resonance imaging
mRNA	messenger ribonucleic acid
NEJM	New England Journal of Medicine
NGF	Neural growth factor
NFL	Neurofilament light chain
NPH	Normal pressure hydrocephalus
NPY	Neuropeptide Y
NRGN	Neurogranin
NSE	Neuron specific enolase
PAI	Plasminogen I activator inhibitor
PC-MR	Phase contrast magnetic resonance
PET	Positron emission tomography
PGDS	Prostaglandin D synthase
PPV	Positive predictive value
QALY	Quality-adjusted life year
R_{CSF}	CSF outflow resistance
Re	CSF reabsorption
SINPHONI	Shunt for idiopathic normal pressure hydrocephalus open-label randomized trials

SAD	Subcortical arteriosclerotic dementia
SAS	Subarachnoid space
SOM	Somatostatin
SPECT	Single photon emission computed tomography
SPLIT	Time-spatial labeling inversion pulse
TNF	Tumor necrosis factor
USD	United States dollar
VAS	Ventriculo-atrial shunt
VEGF	Vascular endothelial growth factor
VIP	Vasoactive intestinal peptide
VPlS	Ventriculo-pleural shunt
VPS	Ventriculo-peritoneal shunt

Author details

Fernando Hakim[1*], Daniel Jaramillo-Velásquez[1*†], Martina González[1], Diego F. Gómez[1], Juan F. Ramón[1] and Mateo Serrano-Pinzón[2]

1 Neurosurgery Department, Fundación Santa Fe de Bogotá, Bogotá D.C., Colombia

2 Medical School, Pontificia Universidad Javeriana, Bogotá D.C., Colombia

*Address all correspondence to: fhakimd@gmail.com; djaramillomd@gmail.com

†This was the author who did the main review of the available literature and wrote the whole manuscript.

References

[1] Williams MA, Malm J. Diagnosis and treatment of idiopathic normal pressure hydrocephalus. CONTINUUM Lifelong Learning in Neurology. 2016;22(2, Dementi):579–99.

[2] Hakim S, Adams RD. The special clinical problem of symptomatic hydrocephalus with normal cerebrospinal fluid pressure. Observations on cerebrospinal fluid hydrodynamics. Journal of the Neurological Sciences. 1965;2(4):307–27.

[3] Hiraoka K, Meguro K, Mori E. Prevalence of idiopathic normal-pressure hydrocephalus in the elderly population of a Japanese rural community. Neurologia Medico-Chirurgica. 2008;48(5):197–9.

[4] Tanaka N, Yamaguchi S, Ishikawa H, Ishii H, Meguro K. Prevalence of possible idiopathic normal-pressure hydrocephalus in Japan: The Osaki-Tajiri project. Neuroepidemiology. 2009;32(3):171–5.

[5] Iseki C, Kawanami T, Nagasawa H, Wada M, Koyama S, Kikuchi K, et al. Asymptomatic ventriculomegaly with features of idiopathic normal pressure hydrocephalus on MRI (AVIM) in the elderly: A prospective study in a Japanese population. Journal of the Neurological Sciences. 2009;277(1–2):54–7.

[6] Nakashita S, Wada-Isoe K, Uemura Y, Tanaka K, Yamamoto M, Yamawaki M, et al. Clinical assessment and prevalence of parkinsonism in Japanese elderly people. Acta Neurologica Scandinavica. 2016;133(5): 373–9.

[7] Andersson J, Rosell M, Kockum K, Lilja-Lund O, Söderström L, Laurell K. Prevalence of idiopathic normal pressure hydrocephalus: A prospective, population-based study. PLoS ONE. 2019;14(5).

[8] Isaacs AM, Riva-Cambrin J, Yavin D, Hockley A, Pringsheim TM, Jette N, et al. Age-specific global epidemiology of hydrocephalus: systematic review, metanalysis and global birth surveillance. PloS one. 2018;13(10): e0204926.

[9] Oliveira LM, Nitrini R, Román GC. Normal-pressure hydrocephalus: A critical review | Hidrocefalia de pressão normal: Uma revisão crítica. Dementia e Neuropsychologia. 2019;13(2):133–43.

[10] Wilson MH. Monro-Kellie 2.0: The dynamic vascular and venous pathophysiological components of intracranial pressure. Journal of Cerebral Blood Flow and Metabolism. 2016;36(8):1338–50.

[11] Bothwell SW, Janigro D, Patabendige A. Cerebrospinal fluid dynamics and intracranial pressure elevation in neurological diseases. Fluids and Barriers of the CNS. 2019;16(1).

[12] Czosnyka M, Czosnyka Z, Momjian S, Pickard JD. Cerebrospinal fluid dynamics. Physiological Measurement. 2004;25(5).

[13] Kim D-J, Kim H, Kim Y-T, Yoon BC, Czosnyka Z, Park K-W, et al. Thresholds of resistance to CSF outflow in predicting shunt responsiveness. Neurological Research. 2015;37(4):332–40.

[14] Benabid AL, de Rougemont J, Barge M. Cerebral venous pressure, sinus pressure and intracranial pressure. Neuro-Chirurgie. 1974;20(7):623–32.

[15] Lu Y-B, Iandiev I, Hollborn M, Körber N, Ulbricht E, Hirrlinger PG, et al. Reactive glial cells: Increased stiffness correlates with increased intermediate filament expression. FASEB Journal. 2011;25(2):624–31.

[16] Fattahi N, Arani A, Perry A, Meyer F, Manduca A, Glaser K, et al.

MR elastography demonstrates increased brain stiffness in normal pressure hydrocephalus. American Journal of Neuroradiology. 2016;37(3): 462–7.

[17] Daneman R, Prat A. The blood–brain barrier. Cold Spring Harbor perspectives in biology. 2015;7(1): a020412.

[18] Liu Q, Radwanski R, Babadjouni R, Patel A, Hodis DM, Baumbacher P, et al. Experimental chronic cerebral hypoperfusion results in decreased pericyte coverage and increased blood–brain barrier permeability in the corpus callosum. Journal of Cerebral Blood Flow and Metabolism. 2019;39(2):240–50.

[19] Eide PK, Hansson H-A. Blood–brain barrier leakage of blood proteins in idiopathic normal pressure hydrocephalus. Brain Research. 2020;1727.

[20] Rasmussen MK, Mestre H, Nedergaard M. The glymphatic pathway in neurological disorders. The Lancet Neurology. 2018;17(11):1016–24.

[21] Eide PK, Ringstad G. Delayed clearance of cerebrospinal fluid tracer from entorhinal cortex in idiopathic normal pressure hydrocephalus: A glymphatic magnetic resonance imaging study. Journal of Cerebral Blood Flow and Metabolism. 2019;39(7):1355–68.

[22] Ringstad G, Vatnehol SAS, Eide PK. Glymphatic MRI in idiopathic normal pressure hydrocephalus. Brain. 2017;140 (10):2691–705.

[23] Reeves BC, Karimy JK, Kundishora AJ, Mestre H, Cerci HM, Matouk C, et al. Glymphatic System Impairment in Alzheimer's Disease and Idiopathic Normal Pressure Hydrocephalus. Trends in Molecular Medicine. 2020;26(3):285–95.

[24] Iliff JJ, Wang M, Liao Y, Plogg BA, Peng W, Gundersen GA, et al. A paravascular pathway facilitates CSF flow through the brain parenchyma and the clearance of interstitial solutes, including amyloid β. Science Translational Medicine. 2012;4(147).

[25] Hasan-Olive MM, Enger R, Hansson H-A, Nagelhus EA, Eide PK. Loss of perivascular aquaporin-4 in idiopathic normal pressure hydrocephalus. GLIA. 2019;67(1):91–100.

[26] Haj-Yasein NN, Vindedal GF, Eilert-Olsen M, Gundersen GA, Skare Ø, Laake P, et al. Glial-conditional deletion of aquaporin-4 (Aqp4) reduces blood–brain water uptake and confers barrier function on perivascular astrocyte endfeet. Proceedings of the National Academy of Sciences. 2011;108 (43):17815–20.

[27] Vindedal GF, Thoren AE, Jensen V, Klungland A, Zhang Y, Holtzman MJ, et al. Removal of aquaporin-4 from glial and ependymal membranes causes brain water accumulation. Molecular and Cellular Neuroscience. 2016; 77:47–52.

[28] Kazui H, Miyajima M, Mori E, Ishikawa M, Hirai O, Kuwana N, et al. Lumboperitoneal shunt surgery for idiopathic normal pressure hydrocephalus (SINPHONI-2): An open-label randomized trial. The Lancet Neurology. 2015;14(6):585–94.

[29] Thomas G, McGirt MJ, Woodworth GF, Heidler J, Rigamonti D, Hillis AE, et al. Baseline neuropsychological profile and cognitive response to cerebrospinal fluid shunting for idiopathic normal pressure hydrocephalus. Dementia and Geriatric Cognitive Disorders. 2005;20(2–3):163–8.

[30] Factora R, Luciano M. Normal Pressure Hydrocephalus: Diagnosis and New Approaches to Treatment. Clinics in Geriatric Medicine. 2006;22(3):645–57.

[31] Kuriyama N, Miyajima M, Nakajima M, Kurosawa M,

Fukushima W, Watanabe Y, et al. Nationwide hospital-based survey of idiopathic normal pressure hydrocephalus in Japan: Epidemiological and clinical characteristics. Brain and Behavior. 2017;7(3).

[32] Nakajima M, Yamada S, Miyajima M, Ishii K, Kuriyama N, Kazui H, et al. Guidelines for management of idiopathic normal pressure hydrocephalus (Third edition): Endorsed by the Japanese society of normal pressure hydrocephalus. Neurologia Medico-Chirurgica. 2021;61 (2):63–97.

[33] Stolze H, Kuhtz-Buschbeck JP, Drücke H, Jöhnk K, Illert M, Deuschl G. Comparative analysis of the gait disorder of normal pressure hydrocephalus and Parkinson's disease. Journal of Neurology Neurosurgery and Psychiatry. 2001;70(3):289–97.

[34] Miyoshi N, Kazui H, Ogino A, Ishikawa M, Miyake H, Tokunaga H, et al. Association between cognitive impairment and gait disturbance in patients with idiopathic normal pressure hydrocephalus. Dementia and Geriatric Cognitive Disorders. 2005;20 (2–3):71–6.

[35] Ogino A, Kazui H, Miyoshi N, Hashimoto M, Ohkawa S, Tokunaga H, et al. Cognitive impairment in patients with idiopathic normal pressure hydrocephalus. Dementia and Geriatric Cognitive Disorders. 2006;21(2):113–9.

[36] Sakakibara R, Kanda T, Sekido T, Uchiyama T, Awa Y, Ito T, et al. Mechanism of bladder dysfunction in idiopathic normal pressure hydrocephalus. Neurourology and Urodynamics. 2008;27(6):507–10.

[37] Krzastek SC, Bruch WM, Robinson SP, Young HF, Klausner AP. Characterization of lower urinary tract symptoms in patients with idiopathic normal pressure hydrocephalus. Neurourology and Urodynamics. 2017; 36(4):1167–73.

[38] Satzer D, Guillaume DJ. Hearing loss in hydrocephalus: a review, with focus on mechanisms. Neurosurgical Review. 2016;39(1):13–25.

[39] Krauss JK, Regel JP, Droste DW, Orszagh M, Borremans JJ, Vach W. Movement disorders in adult hydrocephalus. Movement Disorders. 1997;12(1):53–60.

[40] Mendes GAS, de Oliveira MF, Pinto FCG. The Timed Up and Go Test as a Diagnostic Criterion in Normal Pressure Hydrocephalus. World Neurosurgery. 2017; 105:456–61.

[41] Gallagher R, Marquez J, Osmotherly P. Gait and Balance Measures Can Identify Change from a Cerebrospinal Fluid Tap Test in Idiopathic Normal Pressure Hydrocephalus. Archives of Physical Medicine and Rehabilitation. 2018;99 (11):2244–50.

[42] Krzastek SC, Robinson SP, Young HF, Klausner AP. Improvement in lower urinary tract symptoms across multiple domains following ventriculoperitoneal shunting for idiopathic normal pressure hydrocephalus. Neurourology and Urodynamics. 2017;36(8):2056–2063.

[43] Torralva T, Roca M, Gleichgerrcht E, Lopez P, Manes F. INECO Frontal Screening (IFS): A brief, sensitive, and specific tool to assess executive functions in dementia–CORRECTED VERSION. Journal of the International Neuropsychological Society. 2009;15(5):777–86.

[44] Feick D, Sickmond J, Liu L, Metellus P, Williams M, Rigamonti D, et al. Sensitivity and predictive value of occupational and physical therapy assessments in the functional evaluation of patients with suspected normal

pressure hydrocephalus. Journal of Rehabilitation Medicine. 2008;40(9): 715–20.

[45] Yamada S, Ishikawa M, Yamamot K. Optimal diagnostic indices for idiopathic normal pressure hydrocephalus based on the 3D quantitative volumetric analysis for the cerebral ventricle and subarachnoid space. American Journal of Neuroradiology. 2015;36(12):2262–9.

[46] Ishii K. Diagnostic imaging of dementia with Lewy bodies, frontotemporal lobar degeneration, and normal pressure hydrocephalus. Japanese Journal of Radiology. 2020;38 (1):64–76.

[47] Hashimoto M, Ishikawa M, Mori E, Kuwana N. Diagnosis of idiopathic normal pressure hydrocephalus is supported by MRI-based scheme: A prospective cohort study. Cerebrospinal Fluid Research. 2010;7.

[48] Lee W, Lee A, Li H, Ong NYX, Keong N, Chen R, et al. Callosal angle in idiopathic normal pressure hydrocephalus: small angular mal-rotations of the coronal plane affect measurement reliability. Neuroradiology. 2021.

[49] Park HY, Kim M, Suh CH, Lee DH, Shim WH, Kim SJ. Diagnostic performance and interobserver agreement of the callosal angle and Evans' index in idiopathic normal pressure hydrocephalus: a systematic review and meta-analysis. European Radiology. 2021.

[50] Greenberg MS. Normal Pressure Hydrocephalus. In: Handbook of Neurosurgery. 9th ed. New York: Thieme Medical Publishers Inc.; 2020. p. 417–26.

[51] Yamada S, Tsuchiya K, Bradley WG, Law M, Winkler ML, Borzage MT, et al. Current and emerging MR imaging techniques for the diagnosis and management of CSF flow disorders: A review of phase-contrast and time-spatial labeling inversion pulse. American Journal of Neuroradiology. 2015;36(4):623–30.

[52] Bradley WG. CSF Flow in the Brain in the Context of Normal Pressure Hydrocephalus. AJNR American journal of neuroradiology. 2015;36(5): 831–8.

[53] Marmarou A, Bergsneider M, Klinge P, Relkin N, Black PMcL. INPH guidelines, part III: The value of supplemental prognostic tests for the preoperative assessment of idiopathic normal-pressure hydrocephalus. Neurosurgery. 2005;57(3 SUPPL.).

[54] Vivas-Buitrago T, Lokossou A, Jusué-Torres I, Pinilla-Monsalve G, Blitz AM, Herzka DA, et al. Aqueductal Cerebrospinal Fluid Stroke Volume Flow in a Rodent Model of Chronic Communicating Hydrocephalus: Establishing a Homogeneous Study Population for Cerebrospinal Fluid Dynamics Exploration. World Neurosurgery. 2019;128: e1118–25.

[55] Bradley Jr. WG, Scalzo D, Queralt J, Nitz WN, Atkinson DJ, Wong P. Normal-pressure hydrocephalus: Evaluation with cerebrospinal fluid flow measurements at MR imaging. Radiology. 1996;198(2):523–9.

[56] Scollato A, Tenenbaum R, Bahl G, Celerini M, Salani B, di Lorenzo N. Changes in aqueductal CSF stroke volume and progression of symptoms in patients with unshunted idiopathic normal pressure hydrocephalus. American Journal of Neuroradiology. 2008;29(1):192–7.

[57] Chrysikopoulos H. Idiopathic normal pressure hydrocephalus: Thoughts on etiology and pathophysiology. Medical Hypotheses. 2009;73(5):718–24.

[58] Greitz D, Wirestam R, Franck A, Nordell B, Thomsen C, Ståhlberg F. Pulsatile brain movement and associated hydrodynamics studied by magnetic resonance phase imaging - The Monro-Kellie doctrine revisited. Neuroradiology. 1992;34(5):370–80.

[59] Ringstad G, Emblem KE, Geier O, Alperin N, Eide PK. Aqueductal stroke volume: Comparisons with intracranial pressure scores in idiopathic normal pressure hydrocephalus. American Journal of Neuroradiology. 2015;36(9): 1623–30.

[60] Grazzini I, Venezia D, Cuneo GL. The role of diffusion tensor imaging in idiopathic normal pressure hydrocephalus: A literature review. Neuroradiology Journal. 2021;34(2): 55–69.

[61] Kockum K, Lilja-Lund O, Larsson E-M, Rosell M, Söderström L, Virhammar J, et al. The idiopathic normal-pressure hydrocephalus Radscale: a radiological scale for structured evaluation. European Journal of Neurology. 2018;25(3):569–76.

[62] Kockum K, Virhammar J, Riklund K, Söderström L, Larsson E-M, Laurell K. Diagnostic accuracy of the iNPH Radscale in idiopathic normal pressure hydrocephalus. PLoS ONE. 2020;15(4).

[63] Gómez-Amarillo DF, Pulido LF, Mejía I, García-Baena C, Cárdenas MF, Gómez LM, et al. Cerebrospinal fluid closing pressure-guided tap test for the diagnosis of idiopathic normal pressure hydrocephalus: A descriptive cross-sectional study. Surgical Neurology International. 2020;11.

[64] Hammer M, Sorensen PS, Gjerris F, Larsen K. Vasopressin in the cerebrospinal fluid of patients with normal pressure hydrocephalus and benign intracranial hypertension. Acta Endocrinologica. 1982;100(2):211–5.

[65] Sørensen PS, Gjerris F, Ibsen S, Bock E. Low cerebrospinal fluid concentration of brain-specific protein D2 in patients with normal pressure hydrocephalus. Journal of the Neurological Sciences. 1983;62(1–3): 59–65.

[66] Yamada N, Iwasa H, Mori S, Kurokawa N, Fujimoto K, Kawashima K, et al. Melatonin Secretion in Normal Pressure Hydrocephalus After Cerebral Aneurysm Rupture—Investigation Before and After Ventriculoperitoneal Shunt—. Neurologia medico-chirurgica. 1991;31 (8):490–7.

[67] Malm J, Kristensen B, Ekstedt J, Adolfsson R, Wester P. CSF monoamine metabolites, cholinesterases and lactate in the adult hydrocephalus syndrome (normal pressure hydrocephalus) related to CSF hydrodynamic parameters. Journal of Neurology, Neurosurgery & Psychiatry. 1991;54(3): 252–9.

[68] Hildebrand J, Moussa Z, Raftopoulos C, Vanhouche J, Laute M-A, Przedborski S. Variations of homovanillic acid levels in ventricular cerebrospinal fluid. Acta neurologica scandinavica. 1992;85(5):340–2.

[69] Spanu G, Santagostino G, Marzatico F, Gaetani P, Silvani V, Rodriguez y Baena R. Idiopathic hydrocephalic dementia in aging brain. The neurosurgical approach. Functional Neurology. 1989;4(3):293–8.

[70] Wikkelsö C, Ekman R, Westergren I, Johansson B. Neuropeptides in cerebrospinal fluid in normal-pressure hydrocephalus and dementia. European Neurology. 1991;31 (2):88–93.

[71] Poca MA, Mataró M, Sahuquillo J, Catalán R, Ibañez J, Galard R. Shunt related changes in somatostatin, neuropeptide Y, and corticotropin

releasing factor concentrations in patients with normal pressure hydrocephalus. Journal of Neurology Neurosurgery and Psychiatry. 2001;70 (3):298–304.

[72] Catalan R, Sahuquillo J, Poca MA, Molins A, Castellanos JM, Galard R. Neuropeptide Y cerebrospinal fluid levels in patients with normal pressure hydrocephalus syndrome. Biological Psychiatry. 1994; 36(1):61–3.

[73] Henning RJ, Sawmiller DR. Vasoactive intestinal peptide: Cardiovascular effects. Cardiovascular Research. 2001;49(1):27–37.

[74] Miyajima M, Nakajima M, Ogino I, Miyata H, Motoi Y, Arai H. Soluble amyloid precursor protein α in the cerebrospinal fluid as a diagnostic and prognostic biomarker for idiopathic normal pressure hydrocephalus. European Journal of Neurology. 2013;20 (2):236–42.

[75] Yang J, Dombrowski SM, Krishnan C, Krajcir N, Deshpande A, El-Khoury S, et al. Vascular endothelial growth factor in the CSF of elderly patients with ventriculomegaly: Variability, periodicity and levels in drainage responders and non-responders. Clinical Neurology and Neurosurgery. 2013;115(9): 1729–34.

[76] Huang H, Yang J, Luciano M, Shriver LP. Longitudinal Metabolite Profiling of Cerebrospinal Fluid in Normal Pressure Hydrocephalus Links Brain Metabolism with Exercise-Induced VEGF Production and Clinical Outcome. Neurochemical Research. 2016;41(7):1713–22.

[77] Sutton LN, Wood JH, Brooks BR, Barrer SJ, Kline M, Cohen SR. Cerebrospinal fluid myelin basic protein in hydrocephalus. Journal of neurosurgery. 1983;59(3):467–70.

[78] Whitaker JN, Lisak RP, Bashir RM, Fitch OH, Seyer JM, Krance R, et al. Immunoreactive myelin basic protein in the cerebrospinal fluid in neurological disorders. Annals of Neurology. 1980;7 (1):58–64.

[79] Damasceno BP. Neuroimaging in normal pressure hydrocephalus | Neuroimagem na hidrocefalia de pressão normal. Dementia e Neuropsychologia. 2015;9(4):350–5.

[80] Kitagaki H, Mori E, Ishii K, Yamaji S, Hirono N, Imamura T. CSF spaces in idiopathic normal pressure hydrocephalus: Morphology and volumetry. American Journal of Neuroradiology. 1998;19(7):1277–84.

[81] Yamashita F, Sasaki M, Takahashi S, Matsuda H, Kudo K, Narumi S, et al. Detection of changes in cerebrospinal fluid space in idiopathic normal pressure hydrocephalus using voxel-based morphometry. Neuroradiology. 2010;52(5):381–6.

[82] Yamashita F, Sasaki M, Saito M, Mori E, Kawaguchi A, Kudo K, et al. Voxel-based morphometry of disproportionate cerebrospinal fluid space distribution for the differential diagnosis of idiopathic normal pressure hydrocephalus. Journal of Neuroimaging. 2014;24(4):359–65.

[83] Halperin JJ, Kurlan R, Schwalb JM, Cusimano MD, Gronseth G, Gloss D. Practice guideline: Idiopathic normal pressure hydrocephalus: Response to shunting and predictors of response. Neurology. 2015;85(23):2063–2071.

[84] del Bigio MR, Cardoso ER, Halliday WC. Neuropathological changes in chronic adult hydrocephalus: Cortical biopsies and autopsy findings. Canadian Journal of Neurological Sciences. 1997;24(2):121–6.

[85] Pyykkö OT, Lumela M, Rummukainen J, Nerg O, Seppälä TT,

Herukka S-K, et al. Cerebrospinal fluid biomarker and brain biopsy findings in idiopathic normal pressure hydrocephalus. PloS one. 2014;9(3): e91974.

[86] Leinonen V, Koivisto AM, Savolainen S, Rummukainen J, Tamminen JN, Tillgren T, et al. Amyloid and tau proteins in cortical brain biopsy and Alzheimer's disease. Annals of Neurology. 2010;68(4):446–53.

[87] Abu-Rumeileh S, Giannini G, Polischi B, Albini-Riccioli L, Milletti D, Oppi F, et al. Revisiting the Cerebrospinal Fluid Biomarker Profile in Idiopathic Normal Pressure Hydrocephalus: The Bologna Pro-Hydro Study. Journal of Alzheimer's Disease. 2019;68(2):723–33.

[88] Portelius E, Olsson B, Höglund K, Cullen NC, Kvartsberg H, Andreasson U, et al. Cerebrospinal fluid neurogranin concentration in neurodegeneration: relation to clinical phenotypes and neuropathology. Acta Neuropathologica. 2018;136(3):363–76.

[89] Mattsson N, Insel PS, Palmqvist S, Portelius E, Zetterberg H, Weiner M, et al. Cerebrospinal fluid tau, neurogranin, and neurofilament light in Alzheimer's disease. EMBO Molecular Medicine. 2016;8(10):1184–96.

[90] Kudo T, Mima T, Hashimoto R, Nakao K, Morihara T, Tanimukai H, et al. Tau protein is a potential biological marker for normal pressure hydrocephalus. Psychiatry and Clinical Neurosciences. 2000;54(2):199–202.

[91] Shinoda M, Hidaka M, Lindqvist E, Söderström S, Matsumae M, Oi S, et al. NGF, NT-3 and Trk C mRNAs, but not TrkA mRNA, are upregulated in the paraventricular structures in experimental hydrocephalus. Child's Nervous System. 2001;17(12):704–12.

[92] Tullberg M, Rosengren L, Blomsterwall E, Karlsson J-E, Wikkelsö C. CSF neurofilament and glial fibrillary acidic protein in normal pressure hydrocephalus. Neurology. 1998;50(4):1122–7.

[93] Marcus J, Honigbaum S, Shroff S, Honke K, Rosenbluth J, Dupree JL. Sulfatide is essential for the maintenance of CNS myelin and axon structure. GLIA. 2006;53(4):372–81.

[94] Lukkarinen H, Tesseur I, Pemberton D, van der Ark P, Timmers M, Slemmon R, et al. Time Trends of Cerebrospinal Fluid Biomarkers of Neurodegeneration in Idiopathic Normal Pressure Hydrocephalus. Journal of Alzheimer's Disease. 2021;80(4):1629–42.

[95] Lleó A, Alcolea D, Martínez-Lage P, Scheltens P, Parnetti L, Poirier J, et al. Longitudinal cerebrospinal fluid biomarker trajectories along the Alzheimer's disease continuum in the BIOMARKAPD study. Alzheimer's and Dementia. 2019;15(6):742–53.

[96] Franz G, Beer R, Kampfl A, Engelhardt K, Schmutzhard E, Ulmer H, et al. Amyloid beta 1–42 and tau in cerebrospinal fluid after severe traumatic brain injury. Neurology. 2003;60(9):1457–61.

[97] Neselius S, Brisby H, Theodorsson A, Blennow K, Zetterberg H, Marcusson J. Csf-biomarkers in olympic boxing: Diagnosis and effects of repetitive head trauma. PLoS ONE. 2012;7(4).

[98] Dastague J. Die Paläopathologie. In: Toellner R, editor. Illustrierte Geschichte der Medizin. Special. Salzburg: Andreas & Andreas; 1986. p. 39-undefined.

[99] Aschoff A, Kremer P, Hashemi B, Kunze S. The scientific history of hydrocephalus and its treatment. Neurosurgical Review. 1999;22(2–3): 67–93.

[100] Torkildsen A. A new palliative operation in cases of inoperable occlusion of the Sylvian aqueduct. Acta Chir Scand. 1939; 82:117–25.

[101] Ommaya A. Subcutaneous reservoir and pump for sterile access to ventricular cerebrospinal fluid. The Lancet. 1963;282(7315):983–4.

[102] Cubillos Alonso G. De la válvula de Hakim a la nueva teoría de la mecánica craneana. Bogotá-Universidad de los Andes, Facultad de Medicina, Ediciones Uniandes; 2009. 240-undefined.

[103] Aschoff A, Kremer P, Hashemi B, Oikonomou J, Hampl J, Jansen B, et al. Epidemiology and costs of hydrocephalus. Zentralblt Neurochir (in press). 1999.

[104] Gmeiner M, Wagner H, van Ouwerkerk WJR, Sardi G, Thomae W, Senker W, et al. Long-Term Outcomes in Ventriculoatrial Shunt Surgery in Patients with Pediatric Hydrocephalus: Retrospective Single-Center Study. World Neurosurgery. 2020;138: e112–8.

[105] Hung AL, Vivas-Buitrago T, Adam A, Lu J, Robison J, Elder BD, et al. Ventriculoatrial versus ventriculoperitoneal shunt complications in idiopathic normal pressure hydrocephalus. Clinical Neurology and Neurosurgery. 2017; 157: 1–6.

[106] Milton CA, Sanders P, Steele PM. Late cardiopulmonary complication of ventriculo-atrial shunt. Lancet. 2001; 358(9293):1608.

[107] Pascual JMS, Prakash UBS. Development of Pulmonary Hypertension After Placement of a Ventriculoatrial Shunt. Mayo Clinic Proceedings. 1993;68(12):1177–82.

[108] Ayan E, Tanriverdi HI, Calıskan T, Senel U, Karaarslan N. Intraabdominal pseudocyst developed after ventriculoperitoneal shunt: A case report. Journal of Clinical and Diagnostic Research. 2015;9(6): PD05–6.

[109] Bloch O, McDermott MW. Lumboperitoneal shunts for the treatment of normal pressure hydrocephalus. Journal of Clinical Neuroscience. 2012;19(8):1107–11.

[110] Liu J-T, Su P-H. The efficacy and limitation of lumboperitoneal shunt in normal pressure hydrocephalus. Clinical Neurology and Neurosurgery. 2020;193.

[111] Kim YH, Lee SW, Kim DH, Lee CH, Kim CH, Sung SK, et al. Case series of ventriculoatrial shunt placement in hybrid room: Reassessment of ventriculoatrial shunt. Korean Journal of Neurotrauma. 2020; 16(2):181–9.

[112] Tudor KI, Tudor M, McCleery J, Car J. Endoscopic third ventriculostomy (ETV) for idiopathic normal pressure hydrocephalus (iNPH). Cochrane Database of Systematic Reviews. 2015;(7).

[113] Wang Z, Zhang Y, Hu F, Ding J, Wang X. Pathogenesis and pathophysiology of idiopathic normal pressure hydrocephalus. CNS Neuroscience and Therapeutics. 2020;26 (12):1230–40.

[114] Klinge P, Hellström P, Tans J, Wikkelsø C, Group O behalf of the E iNPH MS. One-year outcome in the European multicentre study on iNPH. Acta Neurologica Scandinavica [Internet]. 2012 Sep 1;126(3):145–53. Available from: https://doi.org/10.1111/j.1600-0404.2012.01676.x

[115] Nakajima M, Miyajima M, Ogino I, Akiba C, Sugano H, Hara T, et al. Cerebrospinal fluid biomarkers for prognosis of long-term cognitive treatment outcomes in patients with idiopathic normal pressure

hydrocephalus. Journal of the Neurological Sciences. 2015;357(1–2): 88–95.

[116] Liu A, Sankey EW, Jusué-Torres I, Patel MA, Elder BD, Goodwin CR, et al. Clinical outcomes after ventriculoatrial shunting for idiopathic normal pressure hydrocephalus. Clinical Neurology and Neurosurgery. 2016; 143:34–8.

[117] Shaw R, Everingham E, Mahant N, Jacobson E, Owler B. Clinical outcomes in the surgical treatment of idiopathic normal pressure hydrocephalus. Journal of Clinical Neuroscience. 2016; 29:81–6.

[118] Kazui H, Kanemoto H, Yoshiyama K, Kishima H, Suzuki Y, Sato S, et al. Association between high biomarker probability of Alzheimer's disease and improvement of clinical outcomes after shunt surgery in patients with idiopathic normal pressure hydrocephalus. Journal of the Neurological Sciences. 2016; 369:236–41.

[119] Kameda M, Yamada S, Atsuchi M, Kimura T, Kazui H, Miyajima M, et al. Cost-effectiveness analysis of shunt surgery for idiopathic normal pressure hydrocephalus based on the SINPHONI and SINPHONI-2 trials. Acta Neurochirurgica. 2017;159(6):995–1003.

[120] Tullberg M, Persson J, Petersen J, Hellström P, Wikkelsø C, Lundgren-Nilsson Å. Shunt surgery in idiopathic normal pressure hydrocephalus is cost-effective—a cost utility analysis. Acta Neurochirurgica. 2018;160(3):509–18.

[121] Alperin N, Oliu CJ, Bagci AM, Lee SH, Kovanlikaya I, Adams D, et al. Low-dose acetazolamide reverses periventricular white matter hyperintensities in iNPH. Neurology. 2014;82(15):1347–51.

[122] Aschoff A. In-vitro-tests von hydrocephalus-ventilen. Habilitationsschrift Universität Heidelberg; 1994.

[123] Haines SJ, Walters BC. Antibiotic prophylaxis for cerebrospinal fluid shunts: a metanalysis. Neurosurgery. 1994;34(1):87–92.

[124] Langley JM, LeBlanc JC, Drake J, Milner R. Efficacy of antimicrobial prophylaxis in placement of cerebrospinal fluid shunts: meta-analysis. Clinical infectious diseases. 1993;17(1):98–103.

[125] di Rocco C, Marchese E, Velardi F. A survey of the first complication of newly implanted CSF shunt devices for the treatment of nontumoral hydrocephalus. Child's Nervous System. 1994;10(5):321–7.

[126] Faulhauer K. The overdrained hydrocephalus. Clinical manifestations and management. In: Advances and technical standards in Neurosurgery. Springer; 1982. p. 3–24.

[127] Giordan E, Palandri G, Lanzino G, Murad MH, Elder BD. Outcomes and complications of different surgical treatments for idiopathic normal pressure hydrocephalus: A systematic review and meta-analysis. Journal of Neurosurgery. 2019;131(4):1024–36.

[128] Lutz BR, Venkataraman P, Browd SR. New and improved ways to treat hydrocephalus: Pursuit of a smart shunt. Surgical Neurology International. 2013;4(SUPPL1).

Chapter 4

Clinical Diagnosis and Treatment Management of Normal Pressure Hydrocephalus

Hüseyin Yakar

Abstract

Inadequate absorption of cerebrospinal fluid (CSF) at the arachnoid granulation level during circulation results in an increase in CSF in the ventricle and certain neuropsychiatric clinical findings. This syndrome, which often presents with ventricular dilatation, progressive cognitive decline, walking difficulties, and urinary incontinence symptoms in elderly individuals, is called Normal Pressure Hydrocephalus (NPH). It is projected that as people's quality of life improves and their life expectancy rises, more old people would develop this condition. Although a clear clinical triad has been defined, the identification of patients with NPH and the application of effective treatment modalities still pose a number of challenges for neurosurgeons today. However, despite all these difficulties, if diagnosed and treated early, the unusual appearance of these symptoms affecting elderly individuals can be prevented and significant improvements in quality of life can be achieved.

Keywords: normal pressure hydrocephalus, elderly individuals, neurodegenerative diseases, cognitive deficits, early surgical treatment

1. Introduction

The brain and spinal cord are soft and vulnerable structures by their very nature. Cerebrospinal fluid (CSF) is the air cushion of the central nervous system (CNS), which protects the nerve tissue by reducing the speed of the blows to the CNS. 90% of this fluid is produced continuously by specialized cells called choroid plexus in the ventricle and 10% by ependymal cells lining the ventricle surface. CSF, which flows through F. Luschka and F. Magendie into the subarachnoid space to surround the brain and spinal cord, drains into the venous system via arachnoid granulation. Colombian neurosurgeon Hakim et al. [1–3] described a clinic in 1964 characterized by progressive cognitive decline with ventricular dilatation (normal CSF pressure during lumbar puncture), difficulty walking, and urinary incontinence syndrome. Hakim named this syndrome Normal Pressure Hydrocephalus (NPH). Although a clear clinical triad has been defined, there are important differences in the clinical presentation and progression of this syndrome. This situation leads to an increase in the problems related to the diagnosis and treatment of NPH. In fact, although there have been remarkable developments in the field of medicine since NPH was first defined in 1964, the guidelines determining the diagnosis, management, and operation criteria of NPH were first prepared in 2004 to be implemented only in

Japan. Only in 2008 did Ishikawa et al. [4] produce worldwide applicable guidelines for its diagnosis and treatment. The information presented above is the most convincing evidence of the type of dynamic disease we are dealing with. On the other hand, due to reasons such as advancements in health, improved treatment options, increased education level, and conscientious diet, the share of the older population is constantly increasing. It is projected that as people's quality of life improves and their life expectancy rises, more old people would develop this condition. In the light of current information, it is predicted that 20% of the world population will be individuals over 65 years old by 2050. While the world's population has grown 4 times in the last 100 years (1950–2050), the fact that the elderly population will grow 10 times is a significant point that should be highlighted. In this case, it becomes even more important to age healthy and to keep the elderly population active. The most important task of neurologists, neurosurgeons, and psychiatrists in society is to provide early diagnosis and appropriate treatment of patients with NPH in society, especially given the socioeconomic consequences of this disease, particularly the burden of dementia on the individual, their families, and society. This is because it is emphasized that the earlier these patients are diagnosed and treated correctly, the more (most if not all) of the clinical symptoms are reversible.

2. Classification

NPH is divided into two groups as secondary NPH (sNPH) that develops due to decreased resorption of CSF due to inflammation and fibrosis at the arachnoid granulation level caused by subarachnoid hemorrhage, intraventricular hemorrhage, meningitis, or traumatic brain injury, and the second is the idiopathic NPH (iNPH) which does not have a causal disorder. A common feature of both diseases is that they do not contain any obstructions to the flow of CSF within the ventricular system of the brain. iNPH and sNPH do not differ in terms of prognosis. The sole significant clinical difference between them is that sNPH affects people of all ages, whereas iNPH often occurs more in the 60-70s [5, 6].

3. Incidence-prevalence

Epidemiological data on NPH are limited. Furthermore, due to the lack of uniform diagnostic criteria, reports on the incidence and prevalence of this disease, which has a wide clinical range, are partially inconsistent. The annual incidence of NPH is estimated to be between 0. 2 and 5.5 cases per 100,000 individuals. Its prevalence is reported to be 0.003% for persons under 65 years of age and 0. 2% to 2.9% for persons 65 years and older [7, 8]. In an epidemiological study conducted by Jaraj et al. [9], the probable prevalence of iNPH was found to be 0. 2% in people aged 70–79, and 5.9% in people aged 80 and older. Another recent epidemiological study also confirmed the inadequacy of incidence-prevalence reports of NPH [10]. Like other neurodegenerative diseases, the prevalence and incidence of NPH increases in direct proportion to age. In various studies, it was determined that there was no difference between males and females in terms of incidence [11–13].

4. Pathophysiology

It is important to clarify its pathophysiology for reliable diagnosis and treatment of NPH patients. Its pathophysiology is yet unknown, and it differs from other adult hydrocephalus causes. In addition to the fact that pathological alterations change CSF pressure, it is also related to changes in CSF dynamics. The CSF circulation spaces in the brain parenchyma within a rigid cranium work as a dynamic system that continually seeks to adapt to new situations in order to keep the ICP constant. These structures give instantaneous responses to changes in CSF production-absorption, changes in arterial–venous flow to the brain, changes in the compliance of intracranial structures, and changes in intracranial pressure. This process is very important in terms of ensuring the correct functioning of the brain. Cerebral blood flow differs with heart rhythm. The arterial supply is pulsative, whereas the venous flow is non-pulsative, causing temporary rises in CSF pressure. In two ways, the system tries to compensate for this. First, vascular structures can reduce arterial blood flow by changing compliance. The second is that the outflow of CSF increases along the cerebral aqueduct. ICP is attempted to be kept constant thanks to these compensatory mechanisms. The decrease in arterial modulation is first compensated by increased pulsatile CSF flow. However, the progressive increase of the pulsatility amplitude causes large ICP pulsations that determine the "water-hammer" effect. These enhanced vibrations create venous damage in the periventricular region, and the process of pushing the brain against the skull continues to expand the ventricles, resulting in hydrocephalus. As a result, the compensatory mechanisms, that are activated in order to maintain the ICP stable, create pathological changes in neural tissue [14]. In fact, hydrocephalus can be defined as the expansion of the ventricles in response to the reduction of the subarachnoid space in the cerebral tissue. This situation is secondary to the increase in the pressure gradient between the ventricles and the subarachnoid space, known as the transmantle pressure [15]. It is still unclear what triggers the initial reduction in arterial compliance in this process. Ischemia emerging in the white matter surrounding arterioles could explain the insufficiency in autoregulation. The ventricular enlargement causes the arterioles and venules around the ventricle to compress and stretch over time, resulting in poor/insufficient cerebral perfusion [16–19]. Moreover, a strong relationship has been described between impaired cerebral blood flow and NPH. Therefore, clinically, the association of NPH with cerebrovascular disease is frequently encountered. Ischemic changes in cerebral tissue caused by decreased/insufficient perfusion were shown in Cranial MRI. These structural changes detected by neuro-radiological imaging have also been supported by neuropathological studies [20–23]. Vascular changes that occur as a natural consequence of aging in humans may be the triggering mechanism in the reduction of vascular compliance. This may explain the relationship between iNPH and vascular disease [24].

NPH also reduces compliance in large vascular structures such as the superior sagittal sinus [25, 26]. Increased transvenular resistance in the sagittal sinuses has been hypothesized as a factor in the onset of NPH. According to this viewpoint, CSF resorption will be affected by increased transvenular resistance [27, 28]. As a result, none of the proposed theories can adequately explain how NPH develops, what factors trigger it, or how structural alterations occur. Although these presented hypotheses appear to complement one other, the debates about pathogenesis continue.

5. Clinic

Symptoms in NPH have been defined as a "triad". However, having all of the symptoms at the same time is not necessary for diagnosis. The presence of two or more of the key symptoms (even a cardinal clinical symptom) such as apraxia of gait, dementia, and urinary incontinence, as well as bilateral dilatation of the ventricles, is necessary to diagnose the disease. The clinical signs and symptoms of this syndrome are highly diverse. Symptoms of this disease, which has an insidious onset, appear gradually over a period of at least 6 months. The rate and extent of worsening of symptoms vary from one patient to another. Some patients and families are unaware of symptoms until a triggering event, such as surgery, occurs. Careful questioning can clarify the nature of symptom onset.

Decreased cerebral perfusion as a result of ventriculomegaly may be a reason for the classic symptoms of NPH. Neurological signs and symptoms, such as apraxia of walking, are thought to be caused by a combination of mechanical stretching of the periventricular fiber tracts, disruption of brain parenchyma tissue as a result of reduced cerebral blood flow, and periventricular edema [29–34]. Neuro-psychiatric symptoms have been suggested to be associated with brain regions such as the anterior cingulate cortex (ACC) and thalamus [35–37] because it has been determined that there is low perfusion in the anterior cingulate cortex and thalamus in NPH patients. Dysfunction in these regions is effective in the emergence of psychiatric symptoms. Therefore, increased/improved cerebral perfusion and oxygen metabolism from the frontal cortex and thalamus may cause neuropsychiatric and other symptoms in NPH patients after shunt surgery [38, 39]. There are publications reporting that psychiatric symptoms and syndromes occurring in the NPH clinic are related to changes in central neurotransmitter activity [40].

Although any of the main symptoms can present as the initial symptom in the NPH clinic, gait and balance disorders usually occur early and have a substantial impact on the individual's life. Dementia and urinary incontinence are symptoms that progress with the disease, albeit they usually appear at later stages of the disease [41].

6. Gait disorder

As described in many published series and guidelines, gait disturbance is the first clinical symptom that affects almost all patients. Dizziness is a common initial complaint among patients. The instability in NPH is better with the patient's eyes open, but patients still stand on a wide base even with their eyes open. When a patient's walking ability is compromised, it has a detrimental influence on their quality of life. At first, gait and balance disorders may appear to be mild. Patients initially complain of climbing and descending stairs, as well as getting up and sitting in a chair. Parallel to the progression of the disease, the patient's gait pattern deteriorates. Instead of the heel-to-toe gait cycle, which should normally be accomplished by raising the feet, these patients tend to slide their feet on the ground. This way of walking is described as "robotic", "sticky-footed" or "magnetic phenomenon" [42]. The disconnection between the basal ganglia and the frontal cortex during walking, as well as the co-contraction of opposing muscles, is suggested to be the source of this gait pattern, which is usually found in parkinsonism (bradykinetic, magnetic) [43, 44]. In the absence of primary sensorimotor deficits, these patients have a higher level of gait disturbance and impaired postural and locomotor reflexes [45]. Gait apraxia develops with the advent of cognitive disorders in the later stages of the disease, and individuals become unable to walk. If these patients are not diagnosed and treated early, they are eventually confined to a wheelchair.

Extrapyramidal symptoms may occur rarely in patients with NPH, but spasticity, hyperreflexia, and other upper motor neuron signs and lateralizing findings are not common. Since the symptoms are bilateral in NPH, lateralizing findings should alert the clinician to the presence of other neuropsychiatric disorders in the differential diagnosis. To assess diagnosis and prognosis, a standard gait assessment (e.g., Tinetti score, Boon Scale) should be performed both before and after the lumbar puncture (LP). The clinical finding with the highest probability of recovery (more than 85 percent) after shunt surgery is apraxia of walking, which is frequently the first main symptom of the disease [46–48].

7. Cognitive disorder

Cognitive deficit in NPH is basically of the "subcortical" type, which includes memory impairment, psychomotor retardation, and impaired ability to apply/use the acquired knowledge [49, 50]. These cognitive and behavioral disorders accompanying NPH are generally defined as "frontal-subcortical dementia or frontal-subcortical dysfunction" [51, 52]. This term is used to describe a pattern of mental decline marked by a lack of interest (apathy) in one's surroundings and oneself, as well as a lack of inner strength (amotivation) that drives one's activities and behaviors [53, 54]. For this reason, patients have difficulty in performing their daily living activities even at the onset of the disease. In this period, it is possible that an abnormality will not be identified in the psychometric tests that will be done on the patients.

Dementia is the most serious symptom in the clinical triad, as it has a negative impact on patients' work capacity as well as their social functioning. NPH is thought to be the etiological cause of 5% of dementia [55]. Even everyday activities like driving, shopping, and keeping track of appointments are challenging for these patients. There is no single type of dementia since dementia symptoms in NPH span a broad clinical spectrum. Instead, depending on the degree of permanent brain damage that has occurred, there are variable degrees of cognitive alteration. For this reason, it is not a very correct approach to define cognitive disorders that occur in NPH as dementia in the early period. Some patients have no clinical evidence of dementia, only mild or moderate cognitive deficits, and most of these patients respond well to shunt surgery [56, 57]. At least two of the following must be present for cognitive abnormalities in NPH patients to be defined as dementia.

- Psychomotor slowing.
- Attention impairment and concentration reduction.
- Short-term memory impairment (cannot repeat learned information).
- In the late phase of the disease, indifference/indifference to environmental stimuli, decreased desire to speak/not speaking at all, decreased thinking/reasoning ability [58].

Since the Mini-Mental State Test and the DEMTEC Test were designed to evaluate cortical dementias, they are not appropriate for evaluating subcortical frontal lobe deficiencies (cognitive deficits) in NPH [59]. The Stroop test, digit span test, and Rey auditory-verbal learning test can be used instead. However, personality changes, anxiety, depression, psychotic syndromes such as delusions, hallucinations, and aggression may also be seen in NPH patients, as well as obsessive–compulsive disorder, Othello syndrome, and various other cognitive disorders such as theft, and mania [60–63]. Depression can be seen in the NPH clinic, although it is rare. In fact, only a tiny portion of these patients who show clinical signs of depression is really diagnosed with depression. Symptoms such as apathy and bradyphrenia that occur in NPH patients may mimic depression. Differential diagnosis between depression and NPH can be challenging as neuropsychological assessment profiles are similar [64, 65]. Therefore, before being diagnosed with depression, NPH patients should have a thorough psychiatric examination, and therapy should be started if actual depression is present. Again, delirium is not encountered in the NPH clinic, and its presence implies the existence of another disease or pharmacological side effect accompanying the disease [41]. Boon AJ et al. [66] reported that iNPH patients showed severe attention deficits. Although the

NPH clinic contains quite different and complex neuropsychiatric symptoms, the decision to have an early shunt surgery can continue to improve cognitive deficits in approximately 80% of patients with NPH, however, the presence of vascular dementia, Alzheimer's dementia, or comorbid diseases at the same time affects the success of surgical treatment negatively and reduces the recovery rate.

8. Urinary incontinence

Urinary symptoms in NPH may occur as urinary frequency, urgency, or incontinence. The bladder dysfunction of NPH is usually in the form of urinary urgency and this condition is almost always present [67, 68]. These patients have difficulty in preventing bladder emptying [69]. Patients have difficulties keeping urinary continence and may suffer urgency with a few drops of urine leakage before reaching the toilet, even though they are aware of the need to urinate at first. Therefore, nocturia is common in NPH patients. Incontinence or having wet clothes are not characteristic of NPH. True urinary incontinence develops later in the course of the disease. While patients initially suffer from increased urinary frequency, they then develop sudden incontinence and eventually persistent urinary incontinence. Bladder dysfunction is due to stretching of the periventricular nerve fibers and loss of subsequent inhibition (partial) of bladder contractions. Bladder function disorders in NPH are caused by detrusor overactivity due to a lack of central inhibitory control, which can be partial or complete [70]. It is extremely rare for fecal incontinence to occur as a symptom of NPH. Therefore, the presence of fecal incontinence in a patient with NPH should first raise suspicion of another type of neurodegenerative disease in the clinician. If a patient with NPH has fecal incontinence as one of the clinical indicators, it suggests he has severe frontal subcortical dysfunction.

When applied early, a CSF shunt can help about 80% of NPH patients with bladder dysfunction; however, if surgery is done at an advanced stage in the disease, as in other symptoms, the percentage would be no more than 50-60%.

9. Diagnosis

For diagnosis, the physical and neurological examinations, clinical symptoms, neuropsychological and neuroimaging findings should all be evaluated as a whole. For this purpose, the clinician should clearly demonstrate the presence of hydrocephalus and the absence of severe cortical atrophy. All patients with NPH should have enlarged ventricles. Although ventriculomegaly is detected in many neurodegenerative diseases and senile cerebral atrophy, these patients may not have any clinical signs of hydrocephalus. Hence, the terms hydrocephalus and ventriculomegaly are not synonymous. To summarize, not all elderly patients with large ventricles have NPH. Ventriculomegaly makes sense when accompanied by clinical symptoms.

Today, in most cases where neurological symptoms are new, Computerized Brain Tomography (CBT) is often used because it is quick and easy to obtain, or Magnetic Resonance Imaging (MRI) because it provides more detailed information about cerebral anatomy/pathology. Furthermore, high-speed and high-resolution MRI techniques can better define aqueductal stenosis, and MRI phase-contrast techniques show the hyperdynamic aqueductal CSF flow that has been associated with shunt-responsive NPH.

Radiological findings detected by MRI/CBT (**Figure 1**).

- Disproportionate ventricular enlargement to sulcal atrophy with typical rounding of frontal horns.

- Periventricular high-density and/or low-density areas (leukoaraiosis) seen diffusely/locally in the white matter due to the transependymal passage of CSF.

- Thinning and elevation of the corpus callosum [71].

- The Evans index, as determined by dilatation of the third and lateral ventricles without obstruction in the CSF circulation and by MRI or CT, should be at least 0. 3 [72].

- Flow gap in the aqueduct detected in spin-echo sequences and called hyperdynamic aqueduct or jet sign (this should be confirmed by hyperdynamic aqueduct phase-contrast MRI) [73].

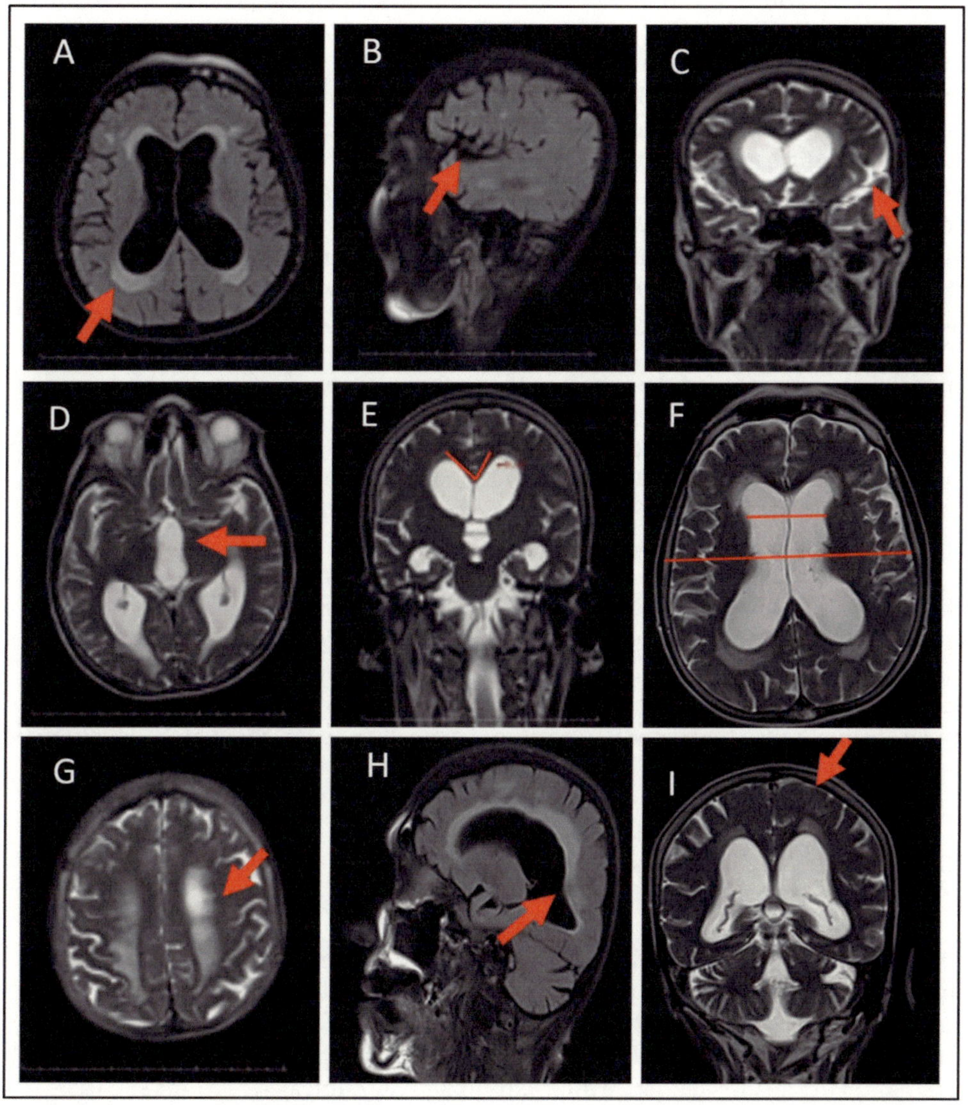

Figure 1.
MRI images of NPH a: Periventricular hyperintensity, B: Enlargement of Sylvian cistern (sagittal), C: Enlargement of Sylvian cistern (coronal), D: Dilatation in the third ventricle, E: Callosal angle, F: Evans index, G: Hyperintensity in white matter, H: Bulging on the roof of the ventricle, I: Effacement of sulci at midline vertex.

The presence of a narrow CSF area in high convexity/midline areas on radiological imaging, and disproportionately enlarged subarachnoid spaces particularly in the Sylvian fissure and basal cisterns, are termed 'Disproportionally Enlarged Subarachnoid Spaces Hydrocephalus' (DESH). This is an indirect sign that CSF flow between the basal cisterns and the arachnoid granulations is being blocked. The existence of this symptom is thought to be the most sensitive indicator for shunt surgery, while its absence indicates brain atrophy [74]. So far, no characteristic neuropathological lesion of NPH has been detected [75–77].

Neuroimaging tests are necessary but not sufficient to diagnose NPH. Invasive tests such as lumbar puncture (LP). and External Lumbar Drainage (ELD) are needed in addition to non-invasive procedures like radiological imaging to improve diagnostic and prognostic accuracy in these patients. Both International and Japanese guidelines recommend diagnostic LP and/or ELD to all patients with suspected NPH. While there is a response to CSF intake in the presence of NPH, there is no response to CSF intake in the absence or minimal level of NPH. CSF drainage also has predictive value for shunt surgery. Patients whose symptoms are relieved by CSF drainage are expected to respond positively to shunt surgery as well. With LP taking 30–50 mL of CSF, changes in gait and cognitive functions are expected after 30 minutes to 4 hours (rarely a few days). If there is a positive response to the tap test, shunt surgery may be recommended, but failure to respond does not exclude the shunt response, because even in patients with normal CSF pressure in the LP, recovery was observed in approximately 50% of them following shunt surgery [78–82]. ELD may be considered in patients who do not respond to the Tap test but are still clinically suspected of having NPH. With ELD, controlled CSF drainage of approximately 10 mL/h for 2–3 days or 150 to 200 mL per day for 2 to 7 days is performed. The patient's gait and neuropsychological tests are recorded daily before the procedure, during CSF drainage, and after catheter removal.

It is difficult to explain the detection of CSF pressure at normal levels in NPH dynamics. Although normal CSF pressure can be detected with a single LP, in fact, 24-hour monitoring might occasionally reveal abnormally high pressures or consistently high/normal pressures. Although CSF pressure has been found to be normal in a single LP, there is a consensus that episodes of increased CSF pressure occur in NPH. For the development of iNPH or sNPH, it is predicted that the baseline ICP is high, at least during the disease stages, and that this high pressure decreases with dilatation of the ventricles. Long-term intracranial pressure (ICP) measurements, such as those taken by some centers for 24 to 72 hours, are not advised for routine usage, both because their predictive values have not yet been adequately documented and because they necessitate specialized equipment and expertise.

10. Differential diagnosis

Regression in motor and cognitive functions, as well as urine incontinence, are common with aging. The addition of other neurodegenerative diseases, such as those that increase with age, and some surgery (cervical/lumbar spinal stenosis) and internal diseases (hypothyroidism, vitamin B12 deficiency) make the differential diagnosis difficult. It may not be easy to distinguish Alzheimer's disease (AD) and Parkinson's disease, which exhibit similar clinical symptoms such as gait disturbance and dementia, from NPH. Also, having vascular or Alzheimer's dementia simultaneously in three-quarters (75%) of their patients with NPH makes the situation even more complicated. On the other hand, because each of the cardinal symptoms of NPH has a variety of etiologies, it might mimic a variety of neurodegenerative diseases. Patients with isolated NPH are extremely uncommon in clinical practice due to the numerous comorbidities that often accompany the symptoms of NPH. The clinical triad peculiar to this disease is actually non-classical, as similar symptoms can be found in a variety of disorders. Therefore, a comprehensive differential diagnosis table ranging from psychiatric disorders to neurological diseases should be considered when distinguishing NPH from other diseases in elderly patients. The differential diagnosis of gait disorders includes peripheral neuropathy, inner ear disorders, spinal cord diseases, alcohol use, and deficiencies of vitamins such as B6 and B12. Clinical and neuroimaging data are very important in the differential diagnosis. Early and accurate determination of the differential diagnosis will save both the clinician and the patient from a series of invasive and noninvasive tests.

Findings that make a diagnosis of NPH less likely include the following:

- ICP: Above 25 cm H_2O.
- AGE: Patients younger than 40 years old.
- SYMPTOM: Asymmetrical or transient symptoms.
- CORTICAL DYSFUNCTION: Having deficits such as aphasia, paresis.
- DEMENTIA: The absence of gait disturbance accompanying the dementia clinic.
- CLINICAL PROCESS: No progression of symptoms.

Some of the diseases frequently encountered in the differential diagnosis are Alzheimer's disease (AD) and Parkinson's disease. Similar to Parkinson's disease, episodes of hesitation and freezing may occur in the gait of NPH patients. However, resting tremors and the typically unilateral symptoms of Parkinson's disease are uncommon in NPH. NPH patients' failure to respond to anti-parkinsonian medicines may also help with diagnosis.

The subject AD, another common disease in differential diagnosis, is quite complex and difficult. AD is thought to account for 50–60% of all dementias in the elderly [83–85]. It is not always possible to distinguish between patients with NPH and those with AD based solely on their medical history and physical examination. Thanks to data gained from MRI and neuropsychological tests, distinguishing AD from NPH is now easier than in past years. The mental disorder in NPH is a subcortical type. While the severity of cognitive impairment is mild or moderate in patients with NPH, mental disorders in AD patients are both the first symptom and advanced. Again, dementia signs occur with more severe symptoms in AD than in NPH. This condition was confirmed by the presence of hippocampal atrophy on CT or MRI [86–89]. Again, motor symptoms such as gait disturbance are rare in AD. In AD, long-term, short-term, and sensory memories are all impaired, while in NPH memory is partially preserved. In NPH, brain dysfunction mainly arises in the frontal cortex, whereas in AD, the major dysfunction originates from the medial temporal lobe, thus, medial temporal lobe atrophy on MRI suggests AD [90]. On the other hand, when considering the response to shunt surgery, it is critical to distinguish these two diseases, which overlap in terms of clinical symptoms. From this standpoint, many studies have investigated biomarkers in CSF to both improve diagnosis and predict shunt efficacy. The specific combination of low Aβ-42 and increased P-tau detected in the CSF has actually been accepted as the biological signature of AD [91]. In contrast, Graff-Radford [92] reported that CSF markers are not useful in distinguishing between the NPH patients from the patients with comorbid AD. Complete blood count, biochemical profile, neuropsychological tests, MRI of the cervical, thoracic or lumbar spine in addition to cranial MRI, electromyography/nerve conduction velocity study and urology consultation can be performed to comprehensively evaluate the differential diagnosis.

11. Treatment

Although NPH is a clinically well-known disease, the indications for shunt surgery and the estimation of surgical outcomes are not clear. Although many devoted articles have been published to identify the most suitable candidates for surgical treatment, there is still no consensus on who is the best candidate for surgery and how to select these patients. Reliable indications of good surgical response are still lacking, particularly with regard to the shunt procedure. In the presence of short history, a known cause of hydrocephalus, predominance of gait disturbances, and CT or MRI findings for hydrodynamic hydrocephalus, it is not difficult to decide on surgery and recommend a shunt to the patient. Today, identifying patients with NPH and applying effective treatment methods still pose challenges for neurosurgeons. However, despite all these difficulties, if diagnosed and treated early, the unusual appearance of these symptoms affecting elderly individuals can be prevented and significant improvements in their life quality can be achieved.

Advanced diagnostic and therapeutic methods and clinical successes have shown that surgical treatment for NPH is superior to conservative treatment. Even if one or two main symptoms are present, NPH should be diagnosed and treated, as waiting for the clinical triad to occur for diagnosis can drastically diminish the response to shunt surgery. This is because the longer NPH patients go without treatment, the worse their prognosis becomes and the shorter their life expectancy becomes.

Using a catheter to alter the flow path of CSF is now the recognized therapeutic procedure all around the world. Shunt surgery is indicated for patients who respond to CSF drainage or who have CSF hydrodynamic variables consistent with NPH [75, 93–95].

However, it is crucial to identify other diseases that mimic NPH before deciding on surgical treatment as it will directly affect the quality of life of patients. There is no evidence that the time spent identifying and treating these disorders in the differential diagnosis lowers the chances of response to shunt surgery. The most essential component that promotes surgical success is a more thorough evaluation performed without haste. Moreover, it should be noted that not all patients with NPH are candidates for shunt surgery. For each patient, the benefit–risk ratio should be assessed separately. Before the surgical operation, possible complications of shunt surgery (infection, embolization, shunt failure, subdural hematoma, and effusion) should be considered and patients should be informed about the surgical risks as well as the potential benefit. Patients should be informed about the problems they will encounter in their daily lives (such as gait disturbance, dementia, incontinence) and potential complications of shunt surgery if they are not operated on. Providing information on the following issues prior to surgical consent will improve the patient's and their relatives' compliance with post-surgery treatment.

a. After surgical treatment, iNPH has a potential cure rate of 30-50% and sNPH of 50-70%.

b. The least reversible symptom with surgical treatment is dementia.

c. The complication rate of surgical treatment varies between 20% and 40%, but serious complications do not exceed 5-8%.

The passage of CSF from one compartment to another by bypassing the natural flow pathways with the aid of a catheter remains the main treatment method for NPH. This shunt procedure is based on the notion that it will minimize the elevated transmantle pressure caused by ventriculomegaly, therefore relieving the symptoms associated with NPH [14]. Today, ventriculoperitoneal (VP) shunts are the most commonly used ones for this purpose. Shunt valves and configuration are dependent on surgeon experience and patient preference. There is no objective evidence that one type of shunt is superior to another. Low-pressure shunts were frequently employed in the past, and the clinical response was better. However, because complications including excessive drainage and subdural hematoma are more common with these shunts, they have been phased out except in rare circumstances. Today, medium pressure shunts or adjustable shunts are more preferred. Adjustable shunts have the advantage of allowing the pressure setting to be gradually lowered or raised until the patient's symptoms improve. In this way, complications that may arise as a result of under or excess drainage can be avoided by changing the pressure without surgery. Another advantage is that it can be administered safely in patients who are on anticoagulation therapy for cardiac or neurological disorders [96].

In Japan, patients with iNPH are mainly treated with lumbar peritoneal shunts. In recent years, this surgical procedure has been widely used all over the world. In terms of effectiveness, one type of shunt has no superiority over the other. However, although the complication rate associated with the device itself is higher in lumbar peritoneal shunts than in ventriculoperitoneal shunts, the fact that lumbar peritoneal shunts are minimally invasive, do not have the fatal complications seen in ventriculoperitoneal shunts, and are more economical has allowed them to be a step forward in treatment [97]. Endoscopic third ventriculostomy has not been proven to be effective in the treatment of iNPH. In patients who are debilitated and shunt surgery is contraindicated, serial lumbar punctures are not recommended as an alternate treatment, except for a limited period of time.

Although it is difficult to draw definitive conclusions, three decades of publications on NPH and surgical experience have summarized the factors that can help predict post-shunt outcomes as follows [98].

a. Factors predicting a good surgical outcome.

- Clinical gait disturbances appearing before cognitive deficits.
- Short duration of mental deterioration history.
- Mild or moderate level of mental disorder.
- Presence of hydrocephalus with known etiology such as subarachnoid hemorrhage, meningitis.
- Detection of significant improvement in clinical findings after CSF drainage.
- Occurrence of 50% or more B waves in continuous intracranial pressure monitoring.
- Absence of significant white matter lesions on MRI.

b. Factors predicting poor surgical outcomes.

- Dementia being the first symptom among clinical findings.
- Detection of clinical signs of severe dementia.
- Detection of significant cerebral atrophy or diffuse white matter involvement on MRI.

Although some studies have indicated a high success (recovery) rate of roughly 80-90% in the improvement of clinical symptoms following surgery [99, 100], the overall rate has been reported to be 65-70% for sNPH cases and 30-50% for iNPH cases [50, 82, 101]. This discrepancy in surgical outcomes could be attributed to the presence of other NPH-related neurodegenerative and/or cerebrovascular disorders. Therefore, meticulousness in differential diagnosis and early treatment of comorbidities can eliminate this inconsistency.

However, the reasons why patients treated with shunts do not respond to shunt surgery are not fully understood. Before concluding that the surgical treatment was unsuccessful, it should be suspected that the failure was due to candidate selection or that the shunt was ineffective in cases where the desired clinical improvement was not achieved after surgery, particularly in patients whose ventricular size did not decrease after shunt or in those who only experienced temporary improvement after surgery [102].

Acknowledgements

I would especially like to thank my colleague İsmail KAYA for his help in English editing of the chapter.

Author details

Hüseyin Yakar
Department of Neurosurgery, Faculty of Medicine, Niğde Ömer Halisdemir University, Niğde, Turkey

*Address all correspondence to: hsyakar@gmail.com

References

[1] Hakim S, Adams RD. The special clinical problem of symptomatic hydrocephalus with normal cerebrospinal fluid pressure. Observations on cerebrospinal fluid hydrodynamics. J Neurol Sci. 1965; 2: 307-327

[2] Hakim S, Venegas JG, Burton JD. The physics of the cranial cavity, hydrocephalus and normal pressure hydrocephalus: mechanical interpretation and mathematical model. Surg Neurol. 1976; 5: 187-210

[3] Adams RD. Fisher CM, Hakim S. et al. Symptomatic occult hydrocephalus with "normal" cerebrospinal-fluid pressure. A treatable syndrome. N Engl J Med. 1965; 273: 117-126

[4] Ishikawa M, Hashimoto M, Kuwana N, Mori E, Miyake H, Wachi A, Takeuchi T, Kazui H, Koyama H. Guidelines for management of idiopathic normal pressure hydrocephalus. Neurologia Medico-Chirurgica. 2008;48 (Suppl):1-23

[5] Torkelson RD, Leibrock LG, Gustavson JL, Sundell RR. Neurological and neuropsychological effects of cerebrospinal fluid shunting in children with assumed arrested "normal pressure" pressure hydrocephalus. J Neurol Neurosurg Psychiatry. 1985;48: 799-806

[6] Bret P, Chazal J. Chronic ("normal pressure") hydrocephalus in childhood and adolescence. A review of 16 cases and reappraisal of the syndrome. Childs Nerv Syst. 1995; 11:687-691

[7] Brean A, Eide PK. Prevalence of probable idiopathic normal pressure hydrocephalus in a Norwegian population. Acta Neurol Scand. 2008; 118:48-53.

[8] Hiraoka K, Meguro K, Mori E. Prevalence of idiopathic normal-pressure hydrocephalus in the elderly population of a Japanese rural community. Neurol Med Chir. 2008; 48: 197-199

[9] Jaraj D, Rabiei K, Marlow T. et al. Prevalence of idiopathic normal-pressure hydrocephalus. Neurology. 2014; 82:1449-1454.

[10] Martin-Laez R, Caballero-Arzapalo H, Lopez-Menendez LA, Arango-Lasprilla JC, Vázquez-Barquero A. Epidemiology of idiopathic normal pressure hydrocephalus: A systematic review of the literature. World Neurosurgery. 2015;87:298-310.

[11] Brean A, Fredo HL, Sollid S, Muller T, Sundstrom T, Eide PK. Five- year incidence of surgery for idiopathic normal pressure hydrocephalus in Norway. Acta Neurol Scand. 2009; 120: 314-316

[12] Tanaka N, Yamaguchi S, Ishikawa H, Ishii H, Meguro K. Prevalence of possible idiopathic normal-pressure hydrocephalus in Japan: the Osaki-Tajiri project. Neuro-epidemiology 2009; 32: 171-175.

[13] Krauss JK, Halve B: Normal pressure hydrocephalus: surveyon contemporary diagnostic algorithms and therapeutic decision making in clinical practice. Acta Neurochir. 2004; 146: 379-388.

[14] Gooriah R, Raman A. Idiopathic normal pressure hydrocephalus: An overview of pathophysiology. Clinical Features, Diagnosis and Treatment. 2016. DOI: 10.5772/64198

[15] Greitz D, Greitz T, Hindmarsh T. A new view on the CSF-circulation with the potential for pharmacological treatment of childhood hydrocephalus. Acta Paediatrica. 1997;86:125-132

[16] Bradley WG. Normal pressure hydrocephalus: New conceptson etiology and diagnosis. AJNR. American Journal of Neuroradiology. 2000; 21:1586-1590

[17] Greitz T. Effect of brain distension on cerebral circulation. Lancet. 1969; 2: 863-865

[18] Sato O, Ohya M, Nojiri K, Tsugane R. Microcirculatory changes in experimental hydrocephalus: morphological and physiological studies. In: Shapiro K,Marmarou A, Portnoy H (eds) Hydrocephalus. Raven, 1984; 215-230

[19] Del Bigio MR, Bruni JE Changes in periventricularvasculature of rabbit brain following induction of hydrocephalu sand after shunting. J Neurosurg. 1988; 69:115-120

[20] Krauss JK, Droste DW, Vach W, Regel JP, Orszagh M,Borremans JJ, Tietz A, Seeger W. Cerebrospinal fluid shunting in idiopathic normal-pressure hydrocephalus of the elderly: Effectof periventricular and deep white matter lesions. Neurosurgery. 1996;39(2):292-300

[21] Tullberg M, Jensen C, Ekholm S, Wikkelsø C. Normal pressure hydrocephalus: Vascular white matter changes on MR images must not exclude patients from shunt surgery. Am J Neuroradiol. 2001;22(9):1665-1673

[22] Akai K, Uchigasaki S, Tanaka U, Komatsu A. Normal pressure hydrocephalus. Neuro-pathological study. Acta Pathologica Japonica. 1987;37(1):97-110

[23] Del Bigio MR. Neuropathological changes caused by hydrocephalus. Acta Neuropathologica. 1993;85(6):573-585

[24] Grünewald RA. Normal pressure hydrocephalus: Pathophysiology. Practical Neurology. 2006;6:264-266

[25] Mase M, Yamada K, Banno T, Miyachi T, Ohara S, Matsumoto T. Quantitative analysis of CSF flow dynamics using MRI in normal pressure hydrocephalus. Acta Neurochirurgica. Supplement. 1998;71:350-353

[26] Bateman GA. Vascular compliance in normal pressure hydrocephalus. Am J Neuroradiol. 2000;21(9):1574-1585

[27] Castro ME, Portnoy HD, Maesaka J. Elevated cortical venous pressure in hydrocephalus. Neurosurgery. 1991;29:232-238

[28] Portnoy HD, Branch C, Castro ME. The relationship of intracranial venous pressure to hydrocephalus. Child's Nervous System. 1994;10:29-35

[29] Oliveira MF, Pinto FCG, Nishikuni K, Botelho RV, Lima AM, Rotta JM. Revisiting hydrocephalus as a model to study brain resilience. Front Hum Neurosci. 2011;5:181.

[30] Greitz TVB, Grepe AOL, Kalmer MSC. Pre- and post-operative evaluation of cerebral blood flow in low pressure hydrocephalus. J Neurosurg. 196931:644-51.

[31] Fisher CM. Hydrocephalus as a cause of gait disturbances in the elderly. Neurology. 1982;32: 1258-1263.

[32] Grubb RL, Raichle ME, Gado MH, Eichling JO, Hughes CP. Cerebral blood flow oxygen utilisation and blood volume in dementia. Neurology. 1977;27:905-910

[33] Meyer JS, Kitagawa Y, Tanahashi N, et al. Pathogenesis of normal pressure hydrocephalus preliminary. Surg Neurol. 1985; 23(2): 121-133

[34] Mathew NT, Meyer JS, Hartmann A, Ott EO. Abnormal cerebrospinalfluid-blood flow dynamics. Implications in diagnosis, treatnent andprognosis in normal pressure hydrocephalus. Arc Neurol. 1975;32:657-664

[35] Craig AH, Cummings JL, Fairbanks L, Itti L, Miller BL, Liand J, Mena I. Cerebral blood flow correlates of apathy in Alzheimer disease, Archives of Neurology. 1996;53: 1116-1120.

[36] Lanctot KL, Moosa S, Herrmann N, Leibovitch FS, Rothenburg L, Cotter A, Black SE, A SPECT study of apathyin Alzheimer's disease, Dementia and Geriatric Cognitive Disorders. 2007;24: 65-72.

[37] Marshall GA, Monserratt L, Harwood D, Mandelkern M. Cummings JL, Sultzer DL. Positron emission tomography metabolic correlates of apathy in Alzheimer disease, Archives of Neurology. 2007;64 :1015-1020

[38] J. Miyamoto, K. Tatsuzawa, Y. Inoue, Y. Imahori and K.Mineura, Oxygen metabolism changes in patients with idiopathic normal pressure hydrocephalus before and after shunting operation, Acta Neurologica Scandinavica. 2007 116137-116143

[39] Tullberg M, Hellstrom P, Piechnik SK, Starmark JE, Wikkelso C. Impaired wakefulness is associated with reduced anterior cingulate CBF in patients with normal pressure hydrocephalus, Acta Neurologica Scandinavica. 2004;110:322-330

[40] Markianos M, Lafazanos S, Koutsis G, Sfagos C, Seretis A. CSF neurotransmitter metabolites and neuropsychiatric symptomatology inpatients with normal pressure hydrocephalus. Clin Neurol Neurosurg. 2009;111:231-234

[41] Williams MA, Relkin NR. Diagnosis and management of idiopathic normal-pressure hydrocephalus. Neurology Clinical Practice. 2013;3:375-385.

[42] Haan J, Jansen ENH, Oostrom J, Roos RAC Falling spells in normal pressure hydrocephalus: A favourable prognostic sign? Eur Neurol. 1987; 27:216-220

[43] Estanol BV Gait apraxia in communicating hydrocephalus. J Neurol Neurosurg Psychiatry. 1981 ;44: 305-308

[44] Knutsson E, Lying-Tunell U. Gait apraxia in normal-pressure hydrocephalus: patterns of movement and muscle activation. Neurology. 1985; 35: 155-160

[45] Nutt JG, Marsden CD, Thompson PD. Human walking and higher-level gait disorders, particularly in the elderly. Neurology. 1993;43:268-279.

[46] Fisher CM (1982) Hydrocephalus as a cause of disturbances of gait inthe elderly. Neurology 32:1358-1363

[47] Sørensen PS, Jansen EC, Gjerris F. Motor disturbances in normal pressure hydrocephalus. Special reference to stance and gait. ArchNeurol. 1986;43:34-38

[48] Graff-Radford NR, Godersky JC. Normal-pressure hydrocephalus. Onset of gait abnormality before dementia predicts good surgical outcome. Arch Neurol. 1986; 43:940-942.

[49] Thomsen AM, Børgesen SE, Bruhn P, Gjerris F. Prognosis of dementia in normal-pressure hydrocephalus after a shunt operation. Ann Neurol. 1986; 20:304-310

[50] Cummings JL, Benson DF Hydrocephalic dementia. In:Dementia: a clinical approach, 2nd edn. Butterworth Heinemann, Boston. 1992:267-291

[51] Ogino A, Kazui H, Miyoshi N, Hashimoto M, Ohkawa S, Tokunaga H,Ikejiri Y, Takeda M. Cognitive impairment in patients with idiopathic normal pressure hydrocephalus Dementia and Geriatric Cognitive Disorders. 2006;21(2):113-119

[52] Tarnaris A, Toma AK, Pullen E, Chapman MD, Petzold A, CipolottiL, Kitchen ND, Keir G, Lemieux L, Watkins LD. Cognitive, biochemical,and imaging profile of patients suffering from idiopathic normal pressure hydrocephalus. Alzheimer's & Dementia. 2011;7(5):501-508.

[53] Larsson A, Wikkelsö C, Bilting M, Stephensen H. Clinical parametersin 74 consecutive patients shunt operated for normal pressure hydrocephalus. Acta Neurologica Scandinavica. 1991;84(6):475-482.

[54] Mega MS, Cummings JL. Frontal-subcortical circuits and neuropsychiatric disorders. The Journal of Neuropsychiatry and Clinical Neurosciences. 1994;6(4):358-370.

[55] Vanneste JA. Diagnosis and management of normal-pressure hydrocephalus, J Neurol. 2000;247: 5-14.

[56] Fisher CM. Hydrocephalus as a cause of disturbances of gait in the elderly. Neurology. 1982:32:1358-1363

[57] Gerloff C, Pickard JD Normal pressure hydrocephalus. In:Brandt T, Caplan LR, Dichgans J, Diener HC, Kennard C (eds) Neurological disorders– course and treatment. Academic, San Diego. 1996:773-778

[58] Michael K, Andreas U. The Differential Diagnosis and Treatment of Normal-Pressure Hydrocephalus. | Dtsch Arztebl Int 2012; 109 (1-2):15-26

[59] Hellstrom P, Edsbagge M, Blomsterwall E, et al.: Neuropsychological effects of shunt treatment in idiopathic normal pressure hydrocephalus. Neurosurgery. 2008; 63: 527-535.

[60] McIntyre AW, Emsley RA. Shoplifting associated with normal pressure hydrocephalus: report of a case. J Geriatr Psychiatry. Neurol. 1990;3:229-230.

[61] Kwentus JA, Hart RP. Normal pressure hydrocephalus presenting as mania. J Nerv Ment Dis. 1987;175: 500-502.

[62] Bloom KK, Kraft WA. Paranoia - an unusual presentation of hydrocephalus. Am J Phys Med Rehabil. 1998;77:157-159.

[63] Yusim A, Anbarasan D, Bernstein C, et al. Normal pressure hydrocephalus presenting as Othello syndrome: case presentation and review of the literature. Am J Psychiatry 2008;165: 1119-1125.

[64] Rosen H, Swigar ME Depression and normal pressure hydrocephalus.A dilemma in neuropsychiatric differential diagnosis. J Nerv Ment Dis. 1976;163:35-40

[65] Hart RP, Kwentus JA. Psychomotor slowing and subcortical type dysfunction in depression. J Neurol Neurosurg Psychiatry. 1987;50: 1263-1266

[66] Boon AJ, Tans JT, Delwel EJ, Egeler-Peerdeman SM, Hanlo PW, Wurzer HA, Avezaat CJ, de Jong DA, Gooskens RH, Hermans J. Dutch normal pressure hydrocephalus study: Prediction of outcome after shunting by resistance to outflow of cerebrospinal fluid. Journal of Neurosurgery. 1997;87:687-693.

[67] Fisher CM Hydrocephalus as a cause of disturbances of gait in the elderly. Neurology 1982;32:1358-1363

[68] Bret P, Chazal J (L'hydrocéphalie chronique de l'adulte.Neurochirurgie. 1990 ; 36 [Suppl] :1-159

[69] Sakakibara R, Kanda T, Sekido T, et al. Mechanism of bladder dysfunction in idiopathic normal pressure hydrocephalus. Neurourol Urodyn 2008;27:507-510

[70] Sakakibara R, Uchiyama T, Kanda T, Uchida Y, Kishi M, Hattori T. Urinary dysfunction in idiopathic normal pressure hydrocephalus. Brain and Nerve. 2008;60(3):233-239

[71] Ishii K, Kanda T, Harada A, Miyamoto N, Kawaguchi T, Shimada K,Ohkawa S, Uemura T, Yoshikawa T, Mori E. Clinical impact of the callosal angle in the diagnosis of idiopathic normal pressure hydrocephalus. European Radiology. 2008;18(11):2678-2683.

[72] Gyldensted C. Measurements of the normal ventricular system and hemispheric sulci of 100 adults with computed tomography. Neuroradiology.1977;14(4):183-192

[73] Bradley WG Jr. CSF flow in the brain in the context of normal pressure hydrocephalus. American Journal of Neuroradiology.2015;36(5):831-838.

[74] Sasaki M, Honda S, Yuasa T, Iwamura A, Shibata E, Ohba H. Narrow CSF space at high convexity and high midline areas in idiopathic normal pressure hydrocephalus detected by axial and coronal MRI. Neuroradiology. 2008;50(2):117-122

[75] Mori E, Ishikawa M, Kato T, et al. Guidelines for management of idiopathic normal pressure hydrocephalus: second edition.Neurol Med Chir 2012;52:775-809

[76] Hashimoto M, Ishikawa M, Mori E, Kuwana N. Study of INPH on neurological improvement (SINPHONI): diagnosis of idiopathic normal pressure hydrocephalus is supported by MRI-basedscheme: a prospective cohort study. Cerebrospinal Fluid Res. 2010;7:18.

[77] Williams MA, Malm J (2016) Diagnosis and Treatment of Idiopathic Normal Pressure Hydrocephalus. Continuum (Minneap Minn) 22:579-599

[78] Greenberg JO, Shenkin HA, Adam R(1977) Idiopathic normal pressure hydrocephalus- a report of 73 patients. J Neurol Neurosurg Psychiatry. 40:336-341.

[79] Black PMcL Idiopathic normal-pressure hydrocephalus. Results of shunting in 62 patients. J Neurosurg. 1980:53:371-377

[80] Pickard JD (1984) Normal pressure hydrocephalus – to shunt or not to shunt. In: Warlow C, Garfield J (eds) Dilemmas in the management of the neurological patient. Churchill Livingstone, Edinburgh, pp. 207-214

[81] Petersen RC, Mokri B, Laws ER . Surgical treatment of idiopathic hydrocephalus in elderly patients. Neurology. 1985; 35:307-311

[82] Vanneste JAL. Three decades of normal pressure hydrocephalus: are we wiser now? J Neurol Neurosurg Psychiatry. 1994; 57:1021-1025

[83] Ebly EM, Parhad IM, Hogan DB, Fung TS. Prevalence and types of dementia in the very old: results from the Canadian Study of Health and Aging. Neurology. 1994; 44: 1593-1599

[84] Ichinowatari N, Tatsunuma T, Makiya H. Epidemiological study of old age mental disorders in the two rural areas of Japan. Jpn J Psychiatry Neurol. 1987;41: 629-636

[85] Sulkava R, WikstroÈm J, Aromaa A, Rautsalo R, Lehtinen V, LahtelaK, Palo J. Prevalence of severe dementia in Finland. Neurology 1985; 35: 1025-1029

[86] Golomb J, de Leon MJ, George AE et al (Hippocampal atrophy in normal pressure hydrocephalus is associated with severity of cognitive impairment. Neurology. 1993;43 [Suppl] 2: 211-212

[87] George AE, de Leon MJ, Miller J, Kluger A, Smith DC CT diagnostic

features of Alzheimer's disease: importance in the choroidal/ hippocampal fissure complex. Am J Neuroradiol. 1990;11:101-107

[88] Golomb J, de Leon MJ, George AE, Kluger A, Convit A, Rusinek H,de Santi S, Litt A, Foo SH, Ferris SH. Hippocampal atrophy correlates with severe cognitive impairment in elderly patients with suspected normal pressure hydrocephalus. J Neurol Neurosurg Psychiatry. 1994; 57:590-593

[89] Holodny AI, Waxman R, George AE, Rusinek H, Kalnin AJ, de Leon M. MR differential diagnosis of normal-pressure hydrocephalus and Alzheimer disease: significance of peri-hippocampal features. Am J Neuroradiol. 1998;19:813-819

[90] Jack CR Jr, Petersen RC, Brien PC, Tangalos EG. MR-based hippocampal volumetry in the diagnosis of Alzheimer disease. Neurology 1992;42(1):183-188.

[91] Blennow K, Hampel H CSF markers for incipient Alzheimer'sdisease. Lancet Neurol. 2003; 2:605-613.

[92] Graff-Radford NR. Alzheimer CSF biomarkers may be misleading innormal-pressure hydrocephalus. Neurology. 2014;83(17):1573-1575

[93] Relkin N, Marmarou A, Klinge P, Bergsneider M, Black PM. Diagnosing idiopathic normal-pressure hydrocephalus: INPH Guidelines, part II. Neurosurgery. 2005;57:4-16.

[94] Marmarou A, Bergsneider M, Klinge P, Relkin N, Black PM. The value of supplemental prognostic tests for the preoperative assessment of idiopathic normal-pressure hydrocephalus: INPH Guidelines, part III. Neurosurgery. 2005;57:17-28.

[95] Bergsneider M, Black PM, Klinge P, Marmarou A, Relkin N.Surgical management of idiopathic normal-pressure hydrocephalus: INPH Guidelines, part IV. Neurosurgery. 2005;57:29-39.

[96] Goodwin CR, Kharkar S, Wang P, Pujari S, Rigamonti D,Williams MA. Evaluation and treatment of patients with suspected normal pressure hydrocephalus on long term warfarin anticoagulation therapy Neurosurgery. 2007;60:497-501.

[97] Nakajima M, Miyajima M, Ogino I, Sugano H, AkibaC, Domon N, Karagiozov KL, Arai H. Use of external lumbar cerebrospinal fluid drainage and lumboperitoneal shunts with strata NSC valves in idiopathic normal pressure hydrocephalus: A single-center experience. World Neurosurgery. 2015;83(3):387-393.

[98] Klinge P, Hellström P, Tans J, Wikkelsø C. European iNPH multicenter study group. One-year outcome in the European multicentre study on iNPH. Acta Neurologica Scandinavica. 2012;126(3):145-153.

[99] Vanneste JAL. Diagnosis and management of normal-pressure hydrocephalus. Neurol. 2000;247:5-14

[100] Black PML, Ojemann RG, Tzouras A (1985) CSF shunts for dementia, incontinence, and gait disturbance. Clin Neurosurg 32:632-651

[101] Poca MA, Solana E, Martínez-Ricarte FR, Romero M, Gándara D,Sahuquillo J. Idiopathic normal pressure: Results of a prospective cohort of 236 shunted patients. Acta Neurochirurgica. Supplement. 2012;114:247-253.

[102] Williams MA, Razumovsky AY, Hanley DF. Evaluation of shunt function in patients who are never better, or better then worse after shunt surgery for NPH. Acta Neurochir. 1998;71:368-370

Section 3

Obstructive Hydrocephalus

Chapter 5

Lesions at the Foramen of Monro Causing Obstructive Hydrocephalus

Ashish Chugh, Sarang Gotecha, Prashant Punia and Neelesh Kanaskar

Abstract

The foramen of Monro has also been referred to by the name of interventricular foramen. The structures comprising this foramen are the anterior part of the thalamus, the fornix and the choroid plexus. Vital structures surround the foramen, the damage to which can be catastrophic leading to disability either temporary or permanent. In the literature it has been shown that tumors occurring in the area of interventricular foramen are rare and usually cause hydrocephalus. The operative approach depends upon the location of the tumor which can be either in the lateral or the third ventricle. Various pathologies which can lead to foramen of Monro obstruction and obstructive hydrocephalus include colloid cyst, craniopharyngioma, subependymal giant cell astrocytoma [SEGA], Neurocysticercosis, tuberculous meningitis, pituitary macroadenoma, neurocytoma, ventriculitis, multiseptate hydrocephalus, intraventricular hemorrhage, functionally isolated ventricles, choroid plexus tumors, subependymomas and idiopathic foramen of monro stenosis. In this chapter, we will discuss the various lesions at the level of foramen of Monro causing obstructive hydrocephalus and the management and associated complications of these lesions based on their type, clinical picture and their appearance on imaging.

Keywords: Foramen of Monro, interventricular foramen, obstruction, obstructive hydrocephalus, raised intracranial pressure

1. Introduction

The foramen of Monro has also been referred to by the name of interventricular foramen. The first description of this foramen was given by Alexander Monro in the year 1783 and 1797. The authors of that era were of the opinion that the use of nomenclature 'foramen of monro' was incorrect; instead 'interventricular foramen' would be more apt. Their reason was that Monro had interpreted the connection between lateral and the third ventricle in an incorrect way.

The structures comprising this foramen are the anterior part of the thalamus, the fornix and the choroid plexus. Vital structures surround the foramen, the damage to which can be catastrophic leading to disability either temporary or permanent.

The dimension of this foramen is not even 1 centimeter and thus the area present for any operative intervention is very small. It is not an easy task for the surgeons to excise the lesion as well as safeguard the vital surrounding structures at the same time.

In the literature it has been shown that tumors occurring in the area of interventricular foramen are rare and usually cause hydrocephalus. The operative approach depends upon the location of the tumor which can be either in the lateral or the third ventricle [1].

In this chapter, we will discuss the various lesions at the level of foramen of Monro causing obstructive hydrocephalus and the management and associated complications of these lesions based on their type, clinical picture and their appearance on imaging.

2. History and anatomy

2.1 Introduction

Fluid balance in central nervous system is basically maintained by cerebrospinal fluid [CSF], which is derived from blood and secreted by choroid plexus lining ventricles of the brain. CSF plays a major buoyancy role in the mechanical support to central nervous system.

Circulation of CSF through ventricular system of cerebral cortex takes place in such a way that there is free communication between cerebral and spinal subarachnoid compartment. It start from lateral ventricular cavities passing through foramen of Monro, then entering into third ventricle passing down along aqueduct of Sylvius and reaching fourth ventricle. The exit from fourth ventricle takes place through foramen of Luschka and foramen of Magendie to subarachnoid space around brainstem and spinal cord [2].

2.2 History

Foramen of Monro is named after a Scottish physician Alexander Monro Secundus [1733–1817], he was third son of Alexander Monro Primus and Isabella MacDonald. He matriculated at Edinburg University in 1745 and received his medical degree in 1755 [3]. He assisted his father Alexander primus in teaching anatomy who held the chair of anatomy at Edinburg University. Monro Secundus recorded detailed descriptions and illustrations regarding communication between lateral and third ventricle of the brain as well as describing changes seen in hydrocephalus [4].

Monro also made several important observations about cranial cavity with application of physical principles to the intracranial contents. George Kellie former student of Monro also studied about blood volume in human brains and reached the same conclusion as his mentor which is now known as Monro-Kellie hypothesis which states that the sum of volumes of brain parenchyma, CSF, and intracranial blood is constant [5].

2.3 Embryology

Central nervous system starts developing from fourth week of intrauterine life. Neural tube formed shows closure of anterior neuropore by middle of fourth week and posterior neuropore by end of fourth week. Cranial end of neural tube shows three dilated brain vesicles as prosencephalon [forebrain], mesencephalon [midbrain] and rhombencephalon [hindbrain]. The procesencephalon further subdivides into an anterior telencephalon which forms two cerebral hemisphere having

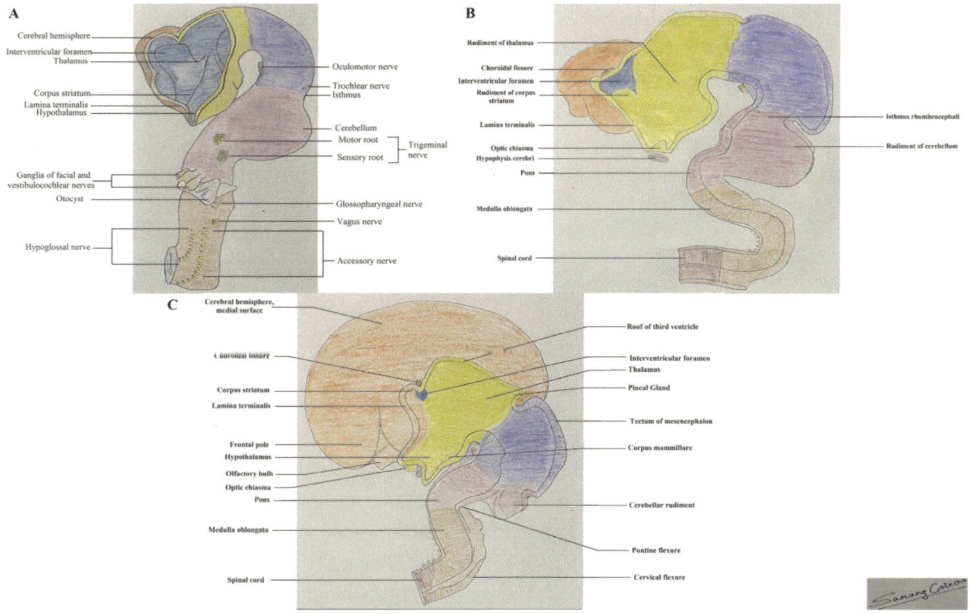

Figure 1.
A] Human embryo, approximately 10.2 mm long, left lateral surface of the diecephlon and telencephalon removed. B] Human embryo 13.6 mm long, median section. C] Human fetus, approximately 3 months old, median section showing medial surface of enlarged telencephalon and diencephalon with interventricular foramen of Monro. [Gray H. THE ANATOMICAL BASIS OF MEDICINE AND SURGERY. 40th ed. Susan Standring, Elsevier Churchill Livingstone New York; 2008. p.381–383.

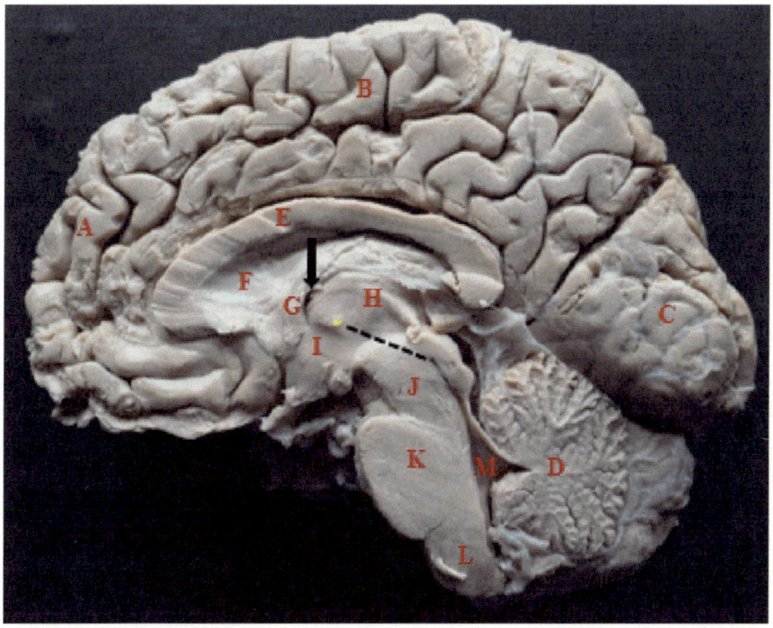

Figure 2.
Sagittal section of brain showing following cerebrum, cerebellum and brainstem:A] frontal lobe, B] Paritel lobe, C] occipital lobe, D] cerebellum, E] corpus callosum, F] septum Pellucidum, G] fornix, H] thalamus, I] hypothalamus, J] pons, K] midbrain, L] medulla oblongata, black arrow- opening of interventricular foramen of Monro, yellow star – Interthalamic adhesion, black dotted line – Thalamohypothalamic sulcus.

lateral ventricle cavities and posterior diencephalon having third ventricle cavity. The commencement of cerebral diverticula from the wall of forebrain persists as interventricular foramen of Monro [6]. Ventricular cavities of brain develop in the neural tube and their enlargement depends of differential growth of the brain vesicles. Out of the three brain vesicles the cavity of the forebrain gives rise to two lateral and third ventricles. Lateral ventricle grows as outpouching from the rostral end of third ventricle and both are interconnected via foramen of Monro (**Figure 1**).

2.4 Gross anatomy

Foramen of Monro is small slit like communicating channel between paired lateral ventricle and third ventricle cavity on either side which become clinically significant when obstructed thus leading to non-communicating hydrocephalus.

Foramen of Monro is located at the junction of roof and anterior wall of lateral ventricle thus bounded anteriorly by the body and column of fornix and posteriorly by anterior nucleus of thalamus (**Figure 2**). Size and shape of the foramen correlates with that of ventricles. In an embryo it is large and circular and as the ventricle size increases, it narrows into a slit like opening. Not only choroid plexus but posterior choroidal, superior choroidal arteries, thalamostriate and septal veins also pass through it (**Figure 2**).

3. Neuro radiology

See **Figure 3**.

3.1 Spectrum of foramen of Monro lesions causing obstructive hydrocephalus (Table 1)

The 3rd ventricle is bounded by the interventricular foramen on either side where the roof and anterior wall of the third ventricle meet the body of the fornix along with the column anterior to the foramen. Posteriorly it is related to the thalamus [anterior pole].

Literature shows that tumors in the vicinity of the foramen are uncommon and usually cause hydrocephalus. Various pathologies which can lead to foramen of Monro obstruction and obstructive hydrocephalus include colloid cyst, craniopharyngioma, subependymal giant cell astrocytoma [SEGA], neurocysticercosis, tuberculous meningitis, pituitary macroadenoma, neurocytoma, ventriculitis, multiseptate hydrocephalus, intraventricular hemorrhage,

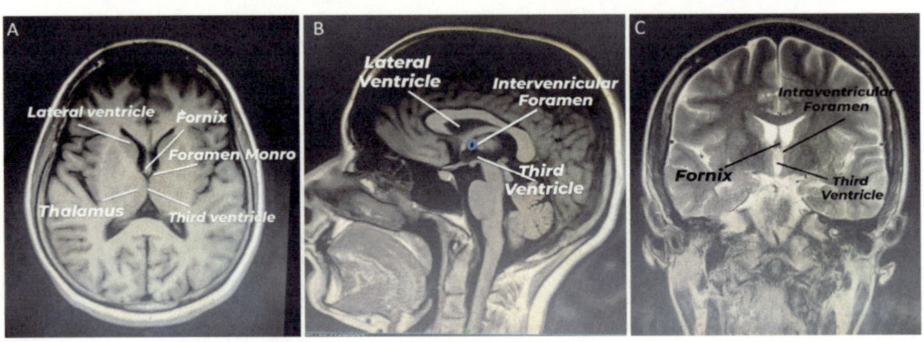

Figure 3.
Normal interventricular foramen and its relations on MRI.

Types of hydrocephalus in foramen of Monro obstruction	Etiological classification	Pathological Classification
• Univentricular • Biventricular • Multiseptate hydrocephalus	• Congenital • Inflammatory • Infective • Neoplastic • Traumatic/ hemorrhagic • Functionally isolated ventricles	• Lesional o IIIrd Ventricle lesion o Lateral ventricle lesion o Septal lesion o Extraventricular lesion • Stenosis Septations/ membranes • Adhesions/exudates • Blood products

Table 1.
Classification of foramen of monro lesions causing obstructive hydrocephalus.

functionally isolated ventricles, choroid plexus tumors, subependymomas and idiopathic foramen of Monro stenosis.

4. Colloid cyst

4.1 Introduction

Colloid cysts are rare, congenital, histologically benign tumors which represent upto 2% of all intracranial neoplasms and correspond to approximately 15–20% of the intraventricular tumors [7]. They occur most commonly in the 3rd to 5th decades of life. They are usually solitary and sporadic, although rare examples of cysts on other locations and familial forms are known. They occur in the posterior end of the foramen of Monro in the anterior and antero-superior part of the third ventricle [8].

The colloid cyst is an epithelial lined cyst filled with gelatinous material which commonly contains mucin, old blood cholesterol and ions and may vary in size from 3 to 40 mm in diameter. However size of the cyst does not appear to be a reliable predictor of outcome, as death may occur even with smaller lesions [9]. Types of colloid material seen can be a] greenish liquid b] greenish liquid with cholesterol c] whitish colloid d] greenish colloid e] mixed.

The precise embryopathogenesis of colloid cyst is poorly understood and still a topic of debate. Various theories proposed are a] origin from either the diencephalic vesicle or the persistence of embryonic paraphysis b] derived from neuroepithelium c] remnant of respiratory epithelium and d] an ependymal cyst from the diencephalon [10].

Clinical presentation is heterogenous and may be intermittent, self- resolving and non-specific. Obstructive hydrocephalus is precipitated by growth of the colloid cyst blocking CSF flow through one or both foramen of Monro and may produce raised intracranial pressure [ICP] by intermittent obstruction of the passage of CSF at the level of interventricular foramen, acting like a legger of a ball as historically described by Dandy in 1933 [8].

4.2 Clinical features

Defining the natural history of colloid cyst reliably has been challenging due to the small number of cases in most case series. Majority of colloid cysts present as an incidental finding while imaging the brain for unrelated symptoms. Symptoms due to colloid cyst often result from different forms of hydrocephalus as well as irritation of major important centers around the third ventricle [11].

They do not have any intrinsic pathological properties and cause symptoms by acting as inert masses. 90% are asymptomatic and stable, while 10% are found to increase in size and cause hydrocephalus. Sudden increase in size can lead to drop attacks, dementia, coma, and death.

Headaches are the most common symptom of colloid cyst ranging from 65–100% of cases according to literature. The headaches are typically severe and intense, throbbing or aching in quality and can be bifrontal or generalized in location and can be precipitated, aggrevated or relieved by head movement or position changes. With progression in size of the cyst, the headaches become more frequent and can be accompanied frequently by nausea, vomiting, blurred vision,

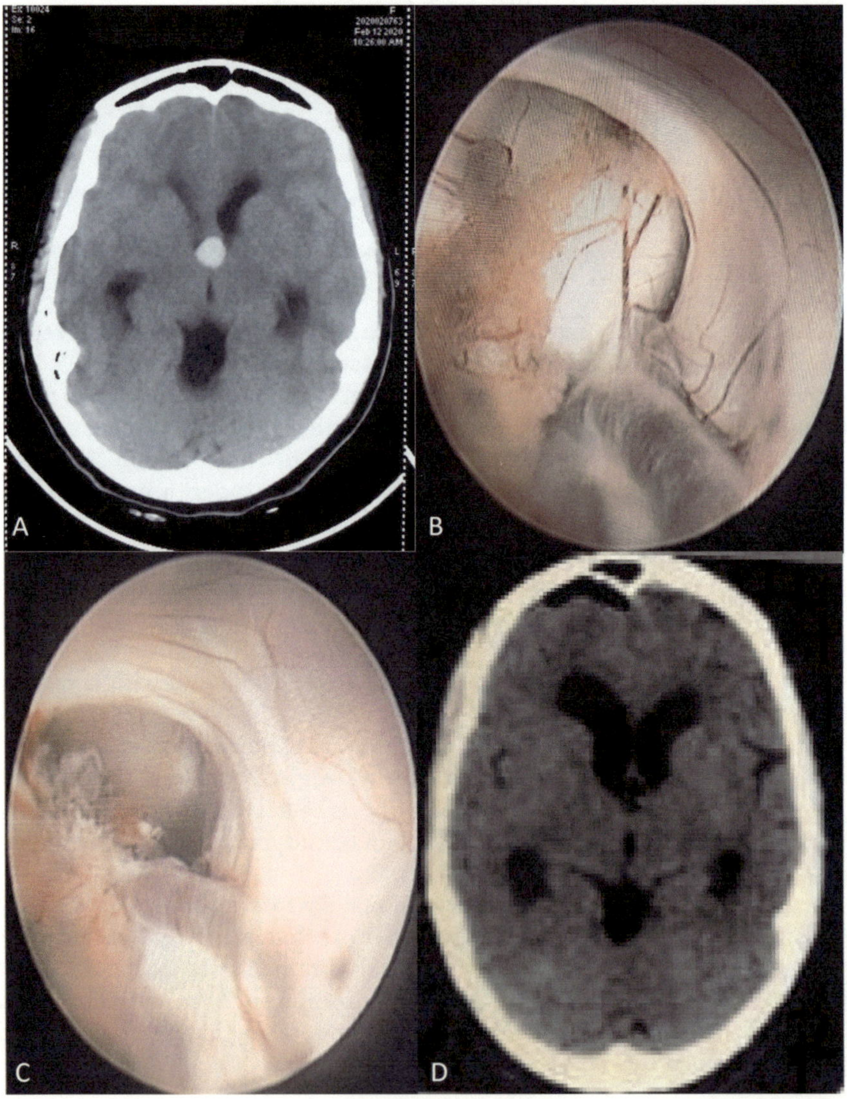

Figure 4.
Case of a small colloid cyst presenting with drop attacks [A-D]: A] pre-op CT showing a hyperdense SOL at foramen of Monro. B] Endoscopic view of the ventricular anatomy and colloid cyst. C] Normal ependyma with 3rd ventricular floor post excision. D] Post op CT showing complete excision of lesion with re-established CSF flow across foramen of monro.

gait ataxia, cognitive decline and less frequently by dizziness, tinnitus and dysautonomic symptoms like abdominal pain, tachycardia, hyperthermia, bradycardia and sweating. Other uncommon symptoms include personality changes, memory disorders, psychiatric disorders and olfactory and gustatory hallucinations. Rapid deterioration due to acute hydrocephalus can occur in 3–35% of patients with an associated 5–38% risk of death [11–13].

4.3 Neuroradiology

Due to the different composition and density of contents, which depends on the quantity of cholesterol and proteins, colloid cysts may have a diverse appearance on imaging. Cysts with high cholesterol and protein content are hyperdense on plain CT, hyperintense on T1 and hypointense on T2 weighted images.

Computed Tomography [CT]: On CT, the colloid cyst is typically a well-defined, round or oval hyperdense mass in the anterior third ventricle at the foramen of Monro. Precontrast scans show a] hyperdense lesions in approximately 2/3rd cases, b] isodense in 1/3rd cases and c] rarely can be hypondense or show calcifications. (**Figure 4**).

On post contrast CT, colloid cysts usually do not show enhancement. Less commonly they may show mild to moderate contrast enhancement. In some cases, thin rim of contrast enhancement may be present which is thought to represent the cyst capsule [8].

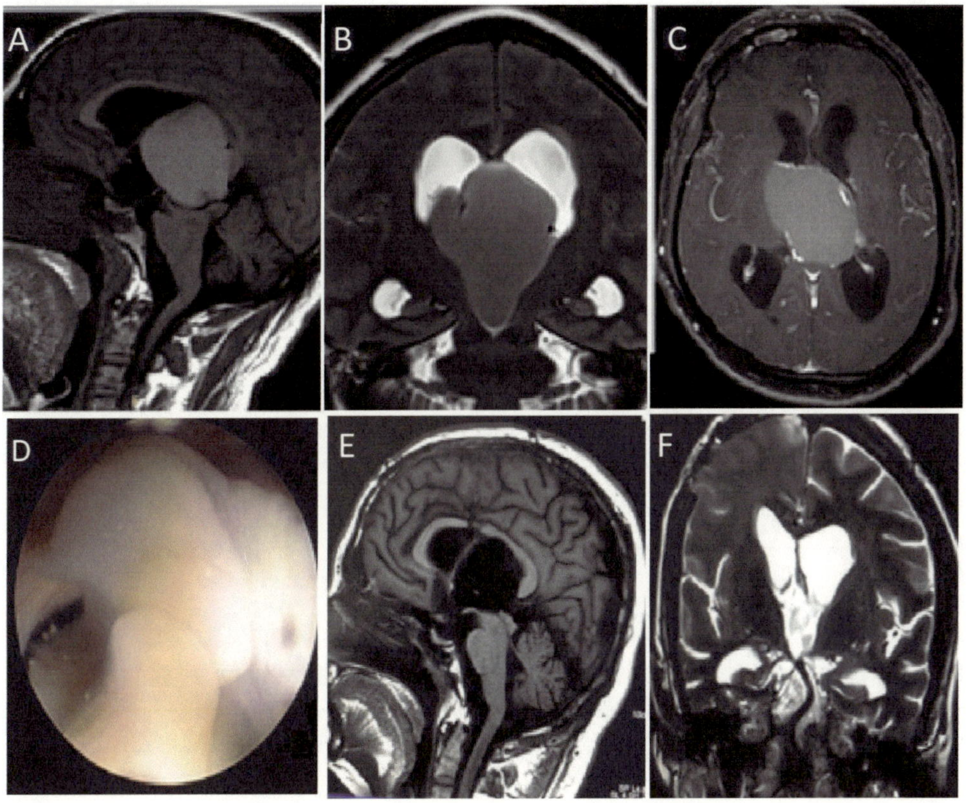

Figure 5.
case of a large colloid cyst presenting with signs of raised ICP A] & B] preoperative Sagittal T1 and Coronal T2 images isodense third ventricular lesion extending into the septum pellucidum. C] preoperative contrast enhanced axial image D] intraoperative image with whitish colloid material being excised E] & F] post-operative T1 sagittal and T2 coronal images showing near total excision.

4.4 Magnetic Resonance Imaging

MRI is the investigation of choice for imaging colloid cysts. On MR imaging, some lesions may show intracystic fluid levels or central and peripheral components in the lesions whereas some lesions are homogenous in appearance.

On T1 Sequencing, colloid cyst is homogenously hyperintense in about 50% of cases while the other lesions can be hyperintense, isointense or hypointense. On T2 Sequence imaging, most colloid cysts are hypointense. Low signal intensity on T2 imaging may correspond to the viscous contents of the colloid cyst and thus harder to aspirate. Post contrast enhancement of the cyst is not seen usually and rarely a peripheral enhancement will be noted around the cyst which represents a vessel stretched over the colloid cyst. (**Figure 5**)

Colloid cysts usually have a similar intensity to surrounding CSF on FLAIR sequencing and may have decreased signal intensity on diffusion weighted imaging [10].

4.5 Treatment

For patients who are symptomatic and have a higher degree of ventriculomegaly, more immediate surgical options include craniotomy for microsurgical resection, neuroendoscopic removal, and CSF -diversion with ventriculo-peritoneal [VP] shunts.

The transcallosal microsurgery and endoscopic approach were compared for the first time by Lewis et al. in 1994. The findings of their comparison were shorter stay in the hospital and early recovery in the endoscopic approach group [14].

In a study by N.B. Levine et al. from 1991 to 2004, the conclusion was made that endoscopy approach should be offered to all patients as the first line of management [15].

Criterion	Points
Age < 65 years	
Yes	1
No	0
Headache	
Yes	1
No	0
Axial diameter ≥ 7 mm	
Yes	1
No	0
FLAIR hyperintensity	
Yes	1
No	0
Risk zone	
Yes	1
No	0

Table 2.
Colloid cyst risk score.

In 2016, Beaumont et al. published the Colloid Cyst Risk Score [CCRS], a method used to stratify the risk of a patient to develop obstructive hydrocephalus and guide physicians to choose appropriate treatment pathways. Risk of lesion progression and obstructive hydrocephalus is more with a CCRS ≥4 [13] (**Table 2**).

In 2016, Suresh Nair et al. published a study which was done from 1980 to 2011, 275 cases were managed by interhemispheric transcallosal approach, twenty-two by transcortical approach and 8 by endoscopic approach. They concluded that open microsurgical approach is the gold standard surgical treatment for colloid cyst [16].

Author recommendations: Neuroendoscopic approach can be offered as the first line treatment for colloid cyst as it can be used for all classic colloid cysts irrespective of the cyst site.

The entry burr hole should be 1 cm lateral to the usual Kocher's point to facilitate endoscopic septostomy. Depending on the intraoperative bleeding, consistency of the cyst and the extent of resection, the decision to place a post-operative EVD can be customized, which can be removed after 4–5 days. Rarely, VP shunt might be needed in these cases.

Case illustrations (**Figures 4** and 5).

5. Craniopharyngioma

5.1 Introduction

Craniopharyngioma [CP] is a benign tumor originating from the squamous epithelial residual cells anywhere along the obscured craniopharyngeal duct from Rathke's cleft to the floor of the third ventricle and they constitute 2–4% of intracranial neoplasms [17]. They often present a surgical challenge to the neurosurgeons because of their central location and proximity to surrounding neural structures namely the hypothalamus, pituitary gland, the optic apparatus, circle of Willis, brainstem and temporal lobes.

Pathologically these tumors are classified into two types: adamantinomous CP [ACP] and papillary CP [PCP]. ACPs are more common than PCPs, are pathologically distinct, are composed of cystic "motor oil-like" component and solid component with frequent calcifications and are more common in pediatric population while PCPs are more common in the adult population [18].

The most common location of craniopharyngioma is the sellar and suprasellar region with 95% of the tumors having a suprasellar component. The suprasellar component of the tumor grows in various directions compressing the surrounding structures. The anatomical proximity of the tumors to the major CSF pathways may result in compression of various parts of the ventricular system causing obstructive hydrocephalus. Obstruction in these cases can be seen at the following levels a] basal cisterns b] invasion and obstruction of the inlet and outlet of the third ventricle c] foramen of Monro and rarely d] posterior displacement of the brainstem with occlusion of the Sylvian aqueduct [19].

Kassam et al. classified suprasellar craniopharyngiomas on the basis of relationship of the tumor to the infundibulum and pituitary stalk as observed in the surgical field on endoscopic viewing which is as follows: I] pre-infundibulum: Its lateral relation are carotid arteries, below by the diaphragm sella, infundibulum in the dorsal portion, and displaced chiasm in the roof. II] trans-infundibulum: The rostral extent of the tumor is bounded by the anterior portion of the hypothalamus. III] retro-infundibulum: a] Extending into the third ventricle. b] Extending into the inter-peduncular cistern. It is bounded anteriorly by the stalk and posteriorly by the mammillary bodies and basilar apex. IV] intra-ventricular and/or optic recess [20].

Generally Kassam's type III and type IV tumors present with obstruction at the level of foramen of Monro.

5.2 Clinical features

Clinical manifestations are related to hypothalamic and pituitary deficiencies, visual impairment and raised intracranial pressure [17].

Headaches are seen in about 50% of patients which may be due to raised intracranial pressure or due to meningeal irritation from the cystic fluid and can be associated with nausea and vomiting.

Symptoms of endocrine dysfunction are seen in 52–87% of patients which are caused by tumor or treatment related lesions to the hypothalamic–pituitary axis that affect the secretion of growth hormone [GH] in about 75% patients, gonadotropins [FSH/LH] in about 40% patients and adrenocorticotropic hormone [ACTH] in about 25% patients. Patients can also present with vasopressin deficiency causing diabetes insipidus in about 20% of cases.

5.3 Investigations

Evaluation and management of craniopharyngioma requires an interdisciplinary approach by endocrinologist, neuro-ophthalmologist and neurosurgeon.

5.4 Neuroradiology

Both CT and MRI are helpful for diagnosis of craniopharyngiomas which are heterogenous tumors with solid, cystic and calcified components.. PCPs are usually solid tumors with rare cystic transformations and calcifications. Radiological features of ACPs can be summarized by the 90% rule i.e. 90% of tumors are predominantly cystic, 90% show more or less prominent calcifications and about 90% take up contrast in the cyst wall [21].

CT provides details of the sellar anatomy as well as information related to cystic and solid components of the tumor, local invasion, compression of adjacent structures and calcifications.

MRI is also useful for the topographic and structural evaluation of craniopharyngiomas. These tumors are isointense or hypointense on T1 weighted images, hypointense or hyperintense on T2 weighted images and show contrast enhancement. This variability in MRI findings are due to varying proportions of solid and cystic components and presence of cholesterol, keratin, hemorrhage and amount of calcification.

5.5 Treatment

The choice of the surgery depends on the anatomical location of the tumor. It has been reported that gross total resection causes more neurological deficits and does not improve the chances of its recurrence. The Endoscopic approach is not recommended for very large tumors with solid component, calcifications or vascular invasion. Kassam's type I and type II tumors can be approached by endoscopic transnasal transsphenoidal technique. However for Kassam's type III and type IV tumors have to be approached by endoscopic transventricular or transcranial approaches (**Table 3**).

The decision regarding the approach to the lesion is guided by the following factors a] consistency of the lesion b] calcification c] lateral and extraventricular extension of the tumor.

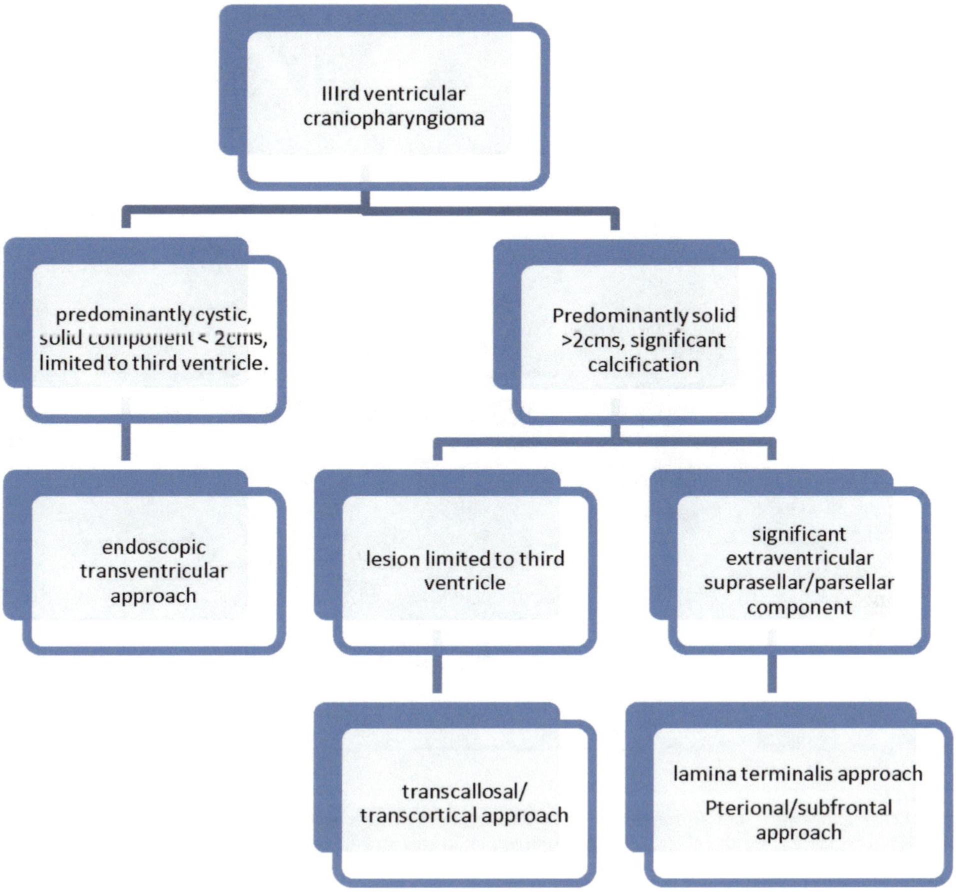

Table 3.
Algorithm for surgical approaches for predominantly third ventricular craniopharygiomas.

Authors recommendations: In the event of endoscopic transventricular approach for craniophryngiomas, the authors strongly recommend endoscopic septostomy and post operative EVD placement in all the cases. VP shunt may be required for permanent CSF drainage depending on the extent of resection. The incidence of postoperative VP shunt at a later date is relatively more common in craniopharyngiomas than colloid cyst. Also Omaya reservoir can be placed post operatively for intralesional bleomycin in predominantly cystic lesions limited to the third ventricle, although the authors have a limited experience regarding the same.

In a study by Deopujari et al. from 2000 to 2016 it was stated that suprasellar craniopharyngiomas with a major cystic component can be best managed by a combined endoscopic transcranial and transnasal approach [22].

Case illustration (**Figure 6**).

The toxic radioactive substances like bleomycin, interferon alpha can be given in craniopharyngiomas with cystic component only to promote sclerosis and fibrosis. One of the major disadvantages of this approach is that it can produce severe neurotoxicity and leakage of the sclerosing substance [23].

For cystic craniopharyngiomas radioisotopes can be implanted through an endoscopic approach. However, there are no standard guidelines for the intra cystic dosage and the toxicity level of these drugs along with tumor control dosage, all these makes this therapeutic option difficult to follow [24].

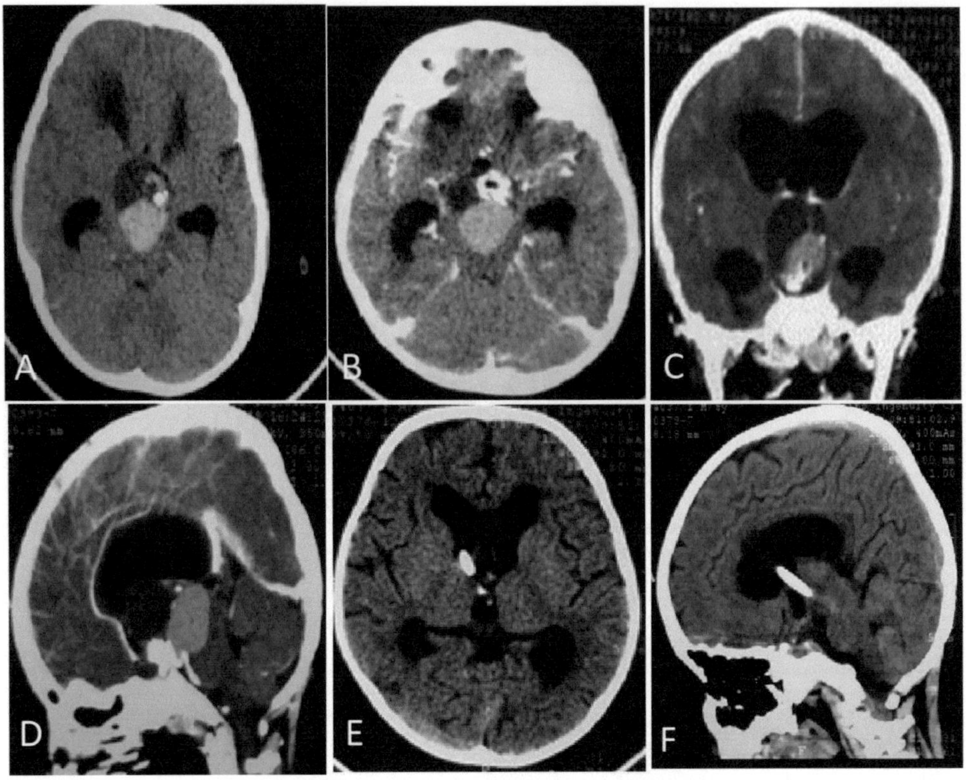

Figure 6.
case of craniopharyngioma [Kassam's type III] presenting with dwarfism and raised ICP A] pre-operative non contrast axial CT B], C] & D] pre-operative contrast enhanced axial, coronal and sagittal CT images respectively E] & F] postoperative non contrast axial and sagittal CT image.

6. Subependymal Giant Cell Astrocytoma [SEGA]

6.1 Introduction

SEGA also known as Subependymal Giant Cell Tumor [SGCT] are clinically benign [WHO grade 1], slow growing tumors which usually arise in the periventricular regions in the proximity of the foramen of Monro. They are associated with tuberous sclerosis complex [TSC], which is a systemic autosomal disease characterized by Vogt's clinical triad of mental retardation, seizures and facial angiofibroma. TSC has a prevalence ranging from 5 to 20% with radiographic evidence of subependymal nodules which are a precursor of SEGA seen in 90% of TSC patients [25, 26].

These are histologically benign tumors and do not undergo malignant transformation. The lesions which are over 10 mm in diameter at the foramen of Monro can cause obstruction of CSF flow leading to progressive dilatation of the lateral ventricles and raised ICP and cause the clinical manifestations associated with SEGA [25].

6.2 Clinical features

Preoperative diagnosis of SEGA takes into account the age, clinical condition of the patients and the location of the tumor. In cases where neurocutaneous manifestations of TSC which include mental retardation, seizures and adenoma sebacum [Vogt's triad] are present, early diagnosis of TSC is possible. However solitary lesions without clinical or radiological features of TSC have also been reported and these patients almost always present urgently due to raised ICP.

In the early stage, patients can present with insidious onset subtle behavior changes, cognitive impairment or seizures. Features of raised ICP like headache and vomiting are present due to increase in size of the lesion causing obstructive hydrocephalus or due to intratumoural or intraventricular hemorrhage [26, 27].

6.3 Investigations

CT and MRI characteristics in SEGA are usually nonspecific. Patient factors which include age and location of the lesion are useful indicators in establishing diagnosis. Although nonspecific, radiological findings show a well circumscribed lesion at the foramen of Monro which is isodense or slightly hyperdense on CT with rare thin calcifications, hypointense on T1, hyperintense on T2 with marked contrast enhancement [26, 28].

6.4 Treatment

SEGAs are considered to be benign lesions and excision of the tumor is considered to be curative in patients presenting with a single lesion. Surgical treatment is indicated in cases of symptomatic SEGA or patients presenting with acute increase

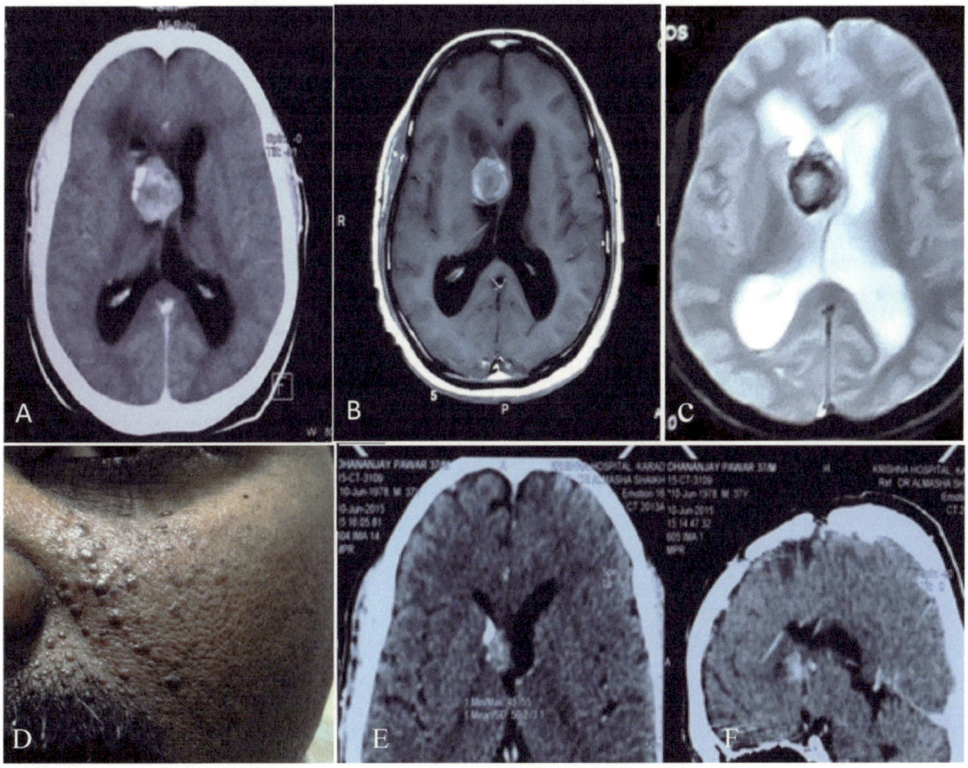

Figure 7.
Young male patient with cutaneous manifestations of TSC presented with sudden onset altered sensorium A] Preoperative CT brain a circumscribed lesion in the right foramen of Monro region with calcification, hemorrhage and obstructive hydrocephalus B] Preoperative MRI showing heteregenous enhancement with mixed cystic and solid areas and multi-stage hemorrhages C] Clinical picture showing cutaneous manifestations of Tuberous Sclerosis Complex D] Follow up CT brain at 2 years showing near total excision of the tumor except the calcified part with resolution of hydrocephalus.

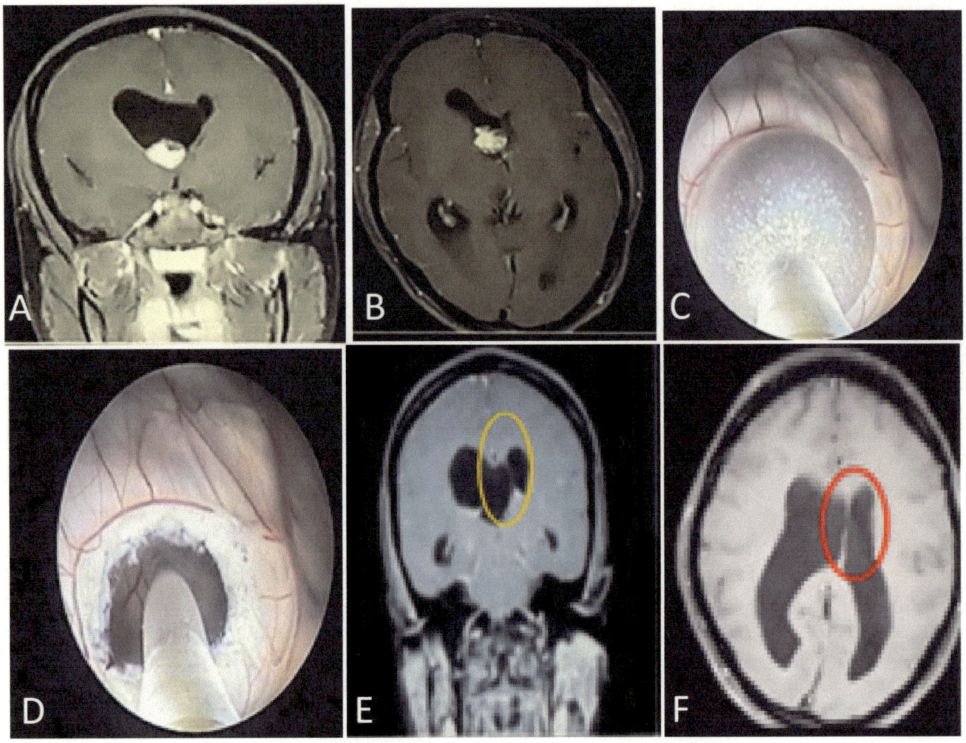

Figure 8.
20 year old female with cuteneous manifestations of TCS presenting with raised ICPA]& B] preoperative MRI showing a contrast enhancing solid lesion at the level of right foramen of Monro leading to univentricular [right lateral ventricle] hydrocephalus. C] & D] endoscopic septostomy was done to deal with the obstructed right ventricle with endoscopic biopsy which was suggestive of SEGA. E] & F] postoperative imaging showing septostomy defect. Later patient was managed with m-TOR inhibitors [rapamycin] with progression free survival over a period of 2 year follow up.

in intracranial pressure due to obstructive hydrocephalus [29]. Other alternative treatment options are Gamma Knife radiosurgery [GKR] or the mechanistic target of rapamycin [mTOR] inhibitors which can reduce the size of the mass in TSC. GKR showed promising results for many types of benign brain tumors, including gliomas, with a low incidence of side effects [30].

Authors recommendations: In our experience, SEGA's were operated by open transcallosal/ transventricular approaches to the lateral ventricle. As these tumors are generally solid and firm in consistency, they are not amenable to endoscopic excision.

Case Illustrations (**Figures 7–9**).

7. Neurocysticerosis

7.1 Introduction

Neurocysticercosis [NCC] occurs when larval stage of the tapeworm Taenia Solium migrates to the central nervous system. It is the most common helminthic infestation of the central nervous system and usually manifests as acute seizure, epilepsy, progressively worsening headache or focal deficit [31].

The cysts reach the ventricular system through the choroid plexus and are more frequently found in the fourth ventricle. This can be attributed to the gravitational

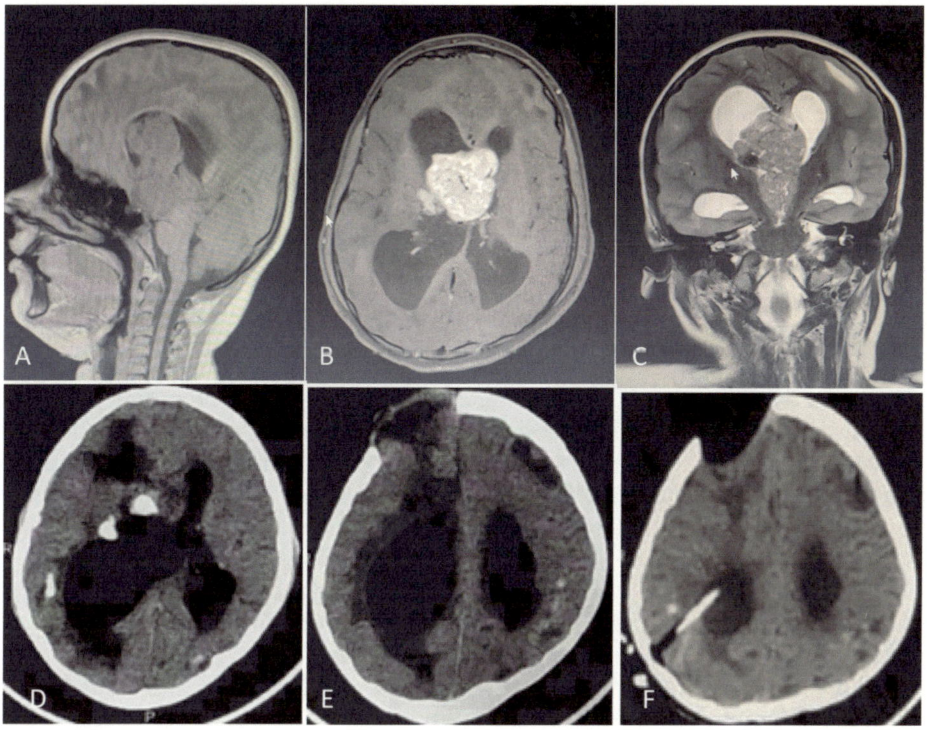

Figure 9.
Mentally retarded child with neurocutaneous markers of TSC presented with raised ICP Figure A], B] & C] MRI revealing large solid enhancing lesion with calcifications at the level of Foramen of monro extending into the lateral ventricle, third ventricle and involving the septum pellucidum causing obstructive hydrocephalus. D], E] & F] Post- operative image showing near total excision with VP shunt in situ [which was done at a later date].

forces which favor migration from supratentorial ventricles or the cysts may directly enter through the choroid plexus [32].

The locations of intraventricular cysts determine the natural history of the disease. Ones attached to the ventricular wall involutes and eventually resolve, while the cysts which are not attached may migrate and block the cerebrospinal flow causing obstructive hydrocephalus. Hydrocephalus can occur either due to ventricular obstruction or arachnoiditis. Sites of obstruction include foramen of Monro, third ventricle, aqueduct of Sylvius and fourth ventricle [32, 33].

7.2 Clinical features

Active intraventricular cysts may remain asymptomatic for years and may become symptomatic if they obstruct the CSF flow leading to hydrocephalus and raised ICP. Abrupt obstruction may lead to acute hydrocephalus and present with features of hydrocephalus which include headache, nausea, vomiting, diplopia, restlessness, seizures, respiratory changes, bradycardia, hypertension, altered sensorium and papilledema or may present infrequently with features of Brun's syndrome [34].

Death of an intraventricular cyst larva can lead to liberation of antigenic substances being released into the ventricular system with local reactions causing an inflammatory response throughout the ventricular system. These patients will present with features of raised ICP, meningoencephalitis, focal neurological deficits and inflammatory reaction detectable in CSF [35].

7.3 Investigations

The diagnosis of intraventricular NCC is based on systematic clinical evaluation, neuroimaging and serology. However the role of serology is limited due to low sensitivity and specificity. Various serological tests include a] antibody testing with immunoblot assay using T. solium antigens on serum or CSF samples b] direct antigen testing of Taenia solium antigen with ELISA using a monoclonal antibody against Taenia saginata HP10 antigens on CSF or serum samples c] detection of Taenia solium specific DNA in CSF by polymerase chain reaction [36].

7.4 Neuroradiology

Identification of scolex in a cystic lesion is the pathognomonic radiological finding in NCC [32, 34, 35]. Non-contrast CT is sensitive for parenchymal and calcified lesions but is not sensitive for extraparenchymal disease. CT may fail to demonstrate small cysts that do not deform the ventricles as a] they share the same density of CSF b] the cyst wall and the scolex are not visible c] the cyst does not show contrast enhancement.

MRI is the investigation of choice for extraparenchymal NCC as the MRI properties of the scolex or the cystic fluid differ and inflammation is marked with hyperintensity signals which allow a reliable diagnosis.

A viable active intraventricular cyst appears as a well defined thin walled cystic lesion of 10–20 mm in diameter which is hypointense on T1 with a thin rim of hyperintensity. The scolex is seen as an eccentric rounded or elongated enhancing mural nodule of 2-4 mm in diameter within the cyst cavity. On T2 weighted imaging, the contents of the cyst are isointense with the surrounding tissues with an hyperintense scolex. Contrast enhanced T1 weighted MRI imaging shows contrast ring enhancement of the cysts with surrounding edema.

7.5 Treatment

Treatment modalities include antiparasitic drugs, surgery and symptomatic medications. Medical management includes a] corticosteroids for meningitis, cysticercal encephalitis and angitis b] antiparasitic drugs which include praziquantel and albendazole [36].

Principles of surgery include treatment of hydrocephalus and removal of cyst. Modalities of surgical intervention include a] emergency ventriculostomy b] placement of VP shunt c] Neuroendoscopic or microsurgical extirpation of obstructing cysts.

As these cysts are not densely adherent to the ventricular wall endoscopic approach is the preferred surgical option. There is usually no enhancement of the cyst wall but in cases where the cyst wall is enhancing, it is suggestive of either ependymal inflammation or adherence of cyst wall to the ependyma. Thus, the decision to approach these lesions endoscopically, in cases of enhancing cyst wall, should be taken with caution. Psarros et al. in his study made an observation that despite rupture of the cyst and spillage of contents in the ventricle, ventriculitis was not seen. Continuous perioperative irrigation with ringers solution helps in removing the debris and provides clear vision [37].

Authors recommendations: Post operative External ventricular drain [EVD] should be placed to address the spillage of contents in the ventricle, but however in our experience we did not find any inflammatory reaction because of the cyst contents.

Case Illustration (**Figure 10**).

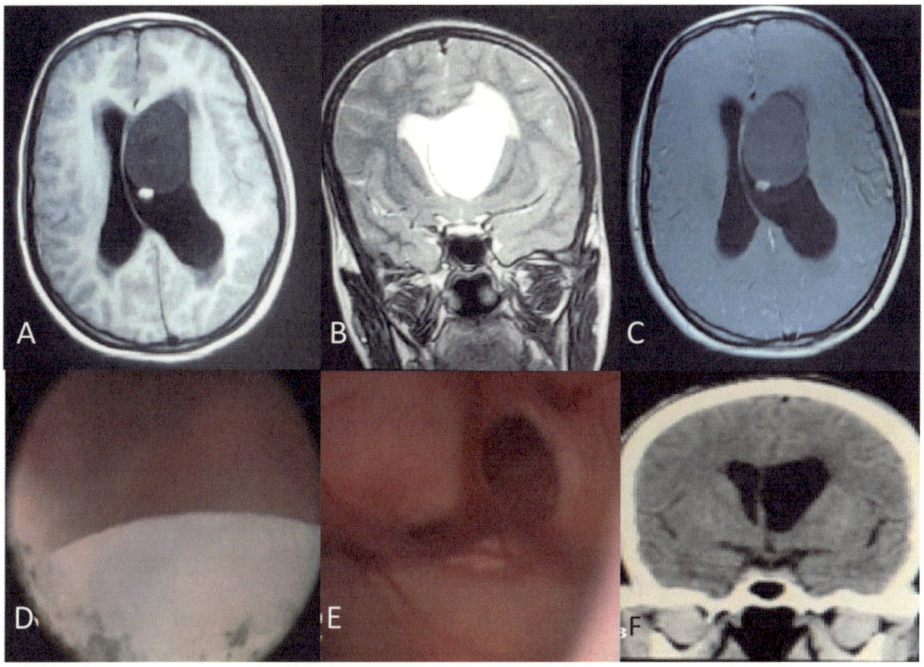

Figure 10.
12 year female presented with Bruns syndrome [episodic symptoms of raised intracranial pressure]. A], B] & C] MRI images showing a cystic lesion in the left lateral ventricle obstructing the left foramen of Monro leading to obstructive hydrocephalus and deviation of the septum pellucidum to the right. D] & E] intraoperative photo showing the excision of the white cyst wall with complete decompression of the left foramen of Monro. F] Postoperative image showing complete excision with midline septum pellucidum.

8. Tubercular meningitis

8.1 Introduction

TBM, a frequent form of central nervous system tuberculosis is a serious neurological disease with significant mortality and morbidity. In India, the estimated mortality due to TBM is approximately 1.5/100,000 population. Neurological complications like cerebral infarctions and hydrocephalus are common and may cause worsening of prognosis [38].

Hydrocephalus, one of the most common complications of TBM, can occur early or late in the clinical course and can also be associated with the commencement of anti-tubercular drugs.

Hydrocephalus in TBM could be either of communicating or obstructive type, the former being more common. The main cause of hydrocephalus in both types is by presence of thick basilar exudates in the subarachnoid spaces or the ventricular pathways [39].

Obstructive hydrocephalus results from block or compression within the fourth ventricle due to exudates or leptomeningeal scar tissue or obstruction of aqueduct of Sylvius due to strangulation of brainstem by exudates or subependymal tuberculoma [40].

Foramen of Monro is anatomically narrowed by the bulk of the choroid plexus which is susceptible to obstruction due to a] meningeal inflammation leading to raised CSF protein content rendering the CSF more viscous and compromising CSF flow b] focal ventriculitis c] exudates/ scarring in the region of foramen of monro [41].

8.2 Clinical features

Progression of disease in patients with obstructive hydrocephalus in TBM is generally rapid as compared to those presenting with communicating hydrocephalus [42].

In addition to the primary and constitutional symptoms of tubercular meningitis, obstructive hydrocephalus should be suspected in any patient with TBM presenting with sudden onset altered sensorium with or without presence of papilledema or in patients complaining of rapidly progressive headache with or without blurring of vision.

8.3 Neuroradiology

Contrast enhanced CT and MRI are helpful in diagnosing the complications of TBM which include presence of hydrocephalus, subependymal seepage, tuberculomas infarcts, edema, nodular enhancing lesions and basal exudates [40].

8.4 Treatment

Medical management includes: antituberculous therapy, steroids, dehydrating agents like mannitol and diuretics such as frusemide and acetazolamide to reduce CSF production [40, 43].

Surgical management includes CSF diversion surgery which can be done conventionally with a VP shunt or neuroendoscopically.

The advantages of neuroendoscopy are:

a. endoscopic septostomy and monroplasty can be done which obviates the need of biventricular shunt

b. provides biopsy sample for further confirmation of the diagnosis

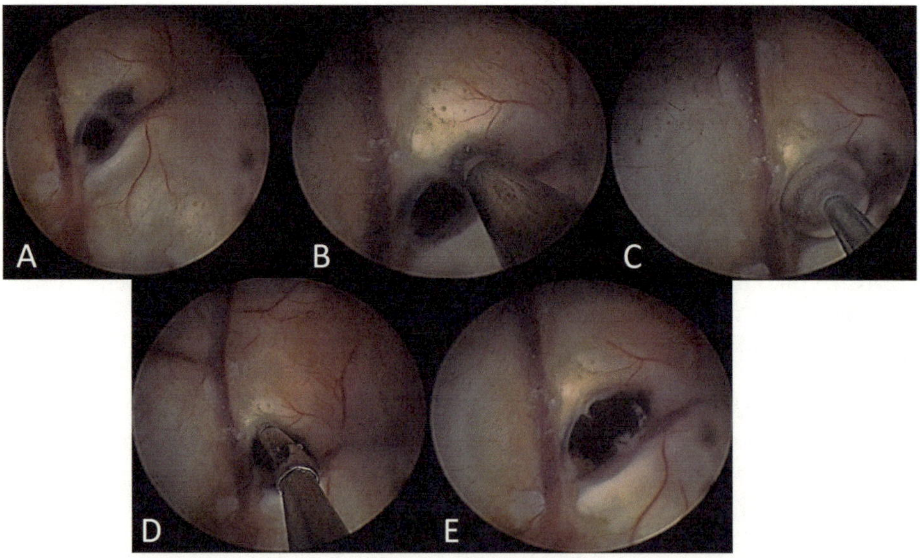

Figure 11.
A] Intra-operative image showing obstruction at the level of foramen of Monro with ependymal tubercles B] perforation of the septum followed by dilatation with C] balloon catheter and D] dilating forceps E] post foraminoplasty restoration of caliber of the foramen of monro with no damage to surrounding structures.

c. adhesiolysis can be performed at the level of foramen of monro

d. clearance of exudates decreases the bulk of the infection and restores the CSF flow which in turn reduces the convalescence period

In a study by Chugh et al. [43]. it was concluded that in cases of TBM with hydrocephalus endoscopic procedures can be offered as the first line of management as it provides the added advantage of performing septostomy, monroplasty and biopsy at the same sitting.

Case Illustration (**Figure 11**).

9. Pituitary adenoma

9.1 Introduction

Pituitary tumors are a heterogenous group of CNS lesions that are usually benign and constitute 10–15% of all primary intracranial tumors in adults. They can be divided on the basis of their size into macradenomas [diameter more than 10 mm] and microadenomas [size less than 10 mm. Nonfunctioning pituitary adenomas with diameter exceeding 40 mm are considered as giant adenomas. These adenomas are rare and frequently invade the suprasellar structures causing mass effect [44, 45].

Obstructive hydrocephalus due to a pituitary adenoma is rare with very less literature available, limited to case reports. Hydrocephalus is rarely the presenting symptom or can occur during the course of the disease.

Various mechanisms of development of hydrocephalus in pituitary adenomas include: a] due to lesser resistance, most pituitary macroadenomas have cranial and anterior extension. The tumor grows along the pituitary stalk upwards through the sellar diaphragm to impinge on the recesses of the third ventricle. In rare cases, the tumor becomes large enough to obstruct the intraventricular foramen leading to secondary hydrocephalus. b] some tumors extend along the floor of the third ventricle and obstruct the Sylvian aqueduct causing obstructive hydrocephalus. c] through obliteration of suprasellar cistern [46].

9.2 Clinical features

Clinical manifestations of giant pituitary adenomas can be secondary to compression of the surrounding structures, pituitary hormone deficiency or hypopituitarism and tumor hypersecretion [44, 46, 47]. Common manifestations of non functioning pituitary adenomas include headache, visual impairment and visual field defects. Obstructive hydrocephalus is characterized by symptoms of raised ICP which include headache, nausea, vomiting, blurring of vision, memory loss, irritability, personality changes, papilledema, sleep disturbances, gait disturbances, loss of bladder control and coma. Involvement of the frontal lobes can be associated with generalized seizures and dementia. Extension of the tumor into the cavernous sinus can be associated with third, fourth and sixth nerve palsy.

9.3 Investigations

Necessary investigations for diagnosing a pituitary tumor include hormonal profile, neuroradiology and fundoscopy.

9.4 Neuroradiology

Contrast enhanced MRI is the investigation of choice for size and location of the tumor and its relation to surrounding structures [48]. CT can provide additional information which include identification of bone destruction and confirmation of suspected intra/parasellar calcifications (**Figure 12**).

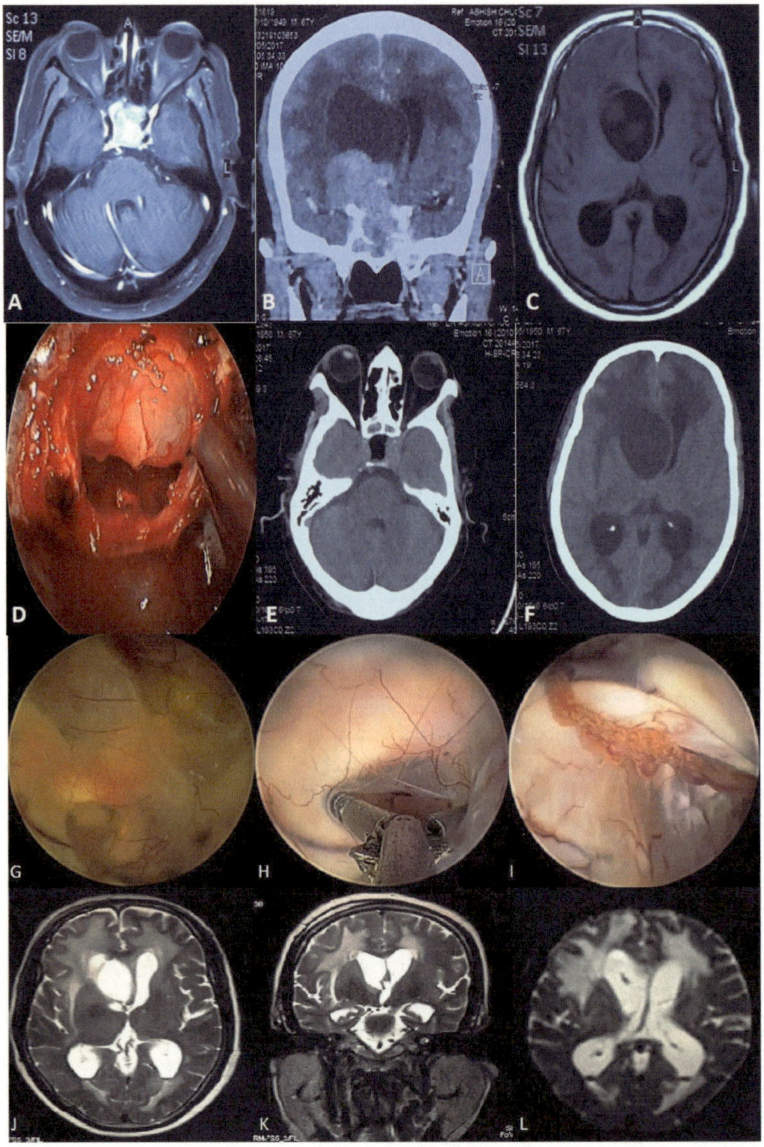

Figure 12.
60 years old male presented with gradual diminution of vision with altered sensorium with A], B] & C] preoperative scans showing a sellar, suprasellar and right parasellar lesion [pituitary adenoma] with a cystic lesion superior to the adenoma causing right foramen of Monro obstruction. D] Intraoperative picture during transnasal transsphenoidal showing complete decompression of the sellar and suprasellar part of the lesion with the diaphragm Sella herniating in the Sella. E], F] postoperative CT showing decompression of the sellar lesion but persistence of the cystic lesion in the right lateral ventricle. G] Neuroendoscopic exploration of the right ventricle shows cystic lesion with xanthochromic fluid with underlying foramen of Monro. H] Post cyst decompression, foraminoplasty being performed. I] Rest of the lateral ventricle examined for remnants of cyst. Also a septostomy was performed and Ommaya reservoir placed post operatively. J], K] & L] post operative scan revealing adequate decompression of the ventricular system with tip of Ommaya at the level of foramen of Monro.

9.5 Management

The goals of surgery include tumor removal, relieving mass effect, improving visual abnormalities, reducing hormone hypersecretion to normal levels and preserving pituitary function [44, 46].

The various modalities of treatment include

a] transsphenoidal approach b] transcranial approach c] combined or two stage approach

In a study by B.K.Ojha et al. it was documented that, a combined trans-sphenoidal and simultaneous transventricular/endoscopic approach is a safe and effective option for giant pituitary macroadenomas in which tumor dimensions, consistency or history of previous surgical treatment were indicative of incomplete removal by single approach alone [49].

Case Illustration.

10. Intraventricular hemorrhage

10.1 Introduction

Intraventricular hemorrhage [IVH] are classified as a] primary – involving the ventricular system and adjacent ventricular lining without associated parenchymal or subarachnoid hemorrhage which occurs in about 30% of cases. b] secondary-primary intracerebral hemorrhage [ICH] or subarachnoid hemorrhage [SAH] extending into the ventricular system which occurs in about 70% of patients and is an independent predictor of poor outcome [50].

Pathophysiology of obstructive hydrocephalus in IVH is due to the blood clot blockage in the CSF pathway. Other contributory factors causing hydrocephalus in these cases include a] release of inflammatory mediators by the blood components causing a secondary response b] damage to the ependymal cells lining the ventricles due to inflammation c] fibrosis and scarring of the arachnoid granulations d] complement activation [51].

10.2 Neuroradiology

CT is the investigation of choice for diagnosing IVH with or without ICH as it allows for rapid diagnosis and prompt management (**Figure 13**). Blood is easily identified on CT as a white hyperdense lesion. CT is also useful to identify other important factors such as edema and hydrocephalus. Pattern and topography of bleeding can give important clues about the secondary causes of IVH and additionally CT angiography or contrast enhanced CT can be done for the same [52].

MRI is equally effective to identify acute hemorrhage. It is useful to distinguish between hemorrhage and an ischemic stroke, although it would be a less preferable investigation in an emergency situation [53].

10.3 Treatment

Goal of treatment is to limit hemorrhagic mass effect, edema, obstructive hydrocephalus by rapid removal of blood and blood products from the ventricular system [54]. Modalities of treatment in these cases include a] EVD insertion b] EVD combined with use of thrombolytics c] Neuroendoscopic aspiration.

Case Illustration (**Figure 13**).

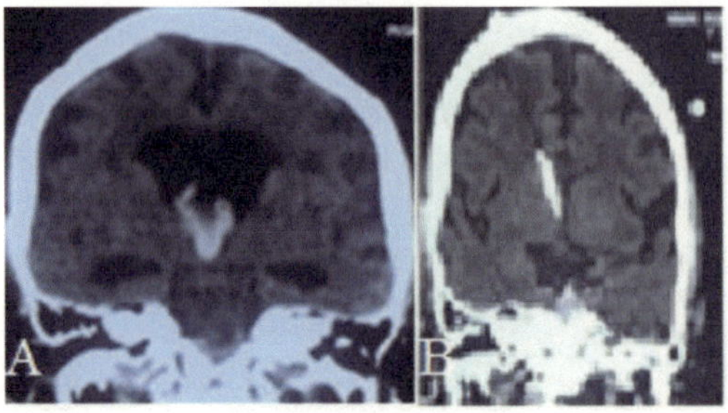

Figure 13.
CT scan A: showing IVH with dilated ventricles. B: Showing EVD in-situ with resolution of IVH and hydrocephalus.

11. Ventriculitis

11.1 Introduction

Nosocomial infection of the central nervous system are a serious complication of patients undergoing neurosurgical procedures like craniotomy, placement of invasive neuromonitoring technique, EVD catheters or CSF shunts.

EVD, which is a common procedure for controlling and monitoring raised ICP secondary to acute occlusive hydrocephalus has an associated major complication of bacterial colonization of the catheter and subsequent retrograde infection resulting in encephalitis, ventriculitis, meningitis, brain abscess, subdural empyema or even sepsis [55].

Ventriculitis, also termed as ependymitis, ventricular empyema, pyocephalus, ventricular abscess or pyogenic ventriculitis, is the inflammation of the ependymal lining of the cerebral ventricles. It can be secondary to meningitis, cerebral abscess, EVD or shunt related, trauma, CSF leak or as a complication of intrathecal chemotherapy.

Incidence of ventriculitis according to literature ranges from 5–20% and incidence of ventricular catheter related ventriculitis ranges from 5–45% [56].

Most common pathogens causing ventriculitis are skin flora which include *staphylococcus epidermidis* [70%] and *staphylococcus aureus* [10%]. Other less common pathogens include gram negative rods [Klebsiella spp., *E. Coli*, Pseudomonas spp], anaerobes and candida spp. [56].

Obstructive hydrocephalus in these cases is due to the obstruction caused by the exudates, adhesions and septations at the level of foramen of Monro or the aqueduct.

11.2 Clinical features

Patients can present with headache, nausea and vomiting, fever, altered mental state, meningism, focal neurological deficits and features of secondary hydrocephalus and raised ICP. On local examination over the subcutaneous shunt tubing, erythema or tenderness can be seen suggestive of infection [57].

11.3 Investigations

Diagnosis of ventriculitis is done by CSF examination and neuroradiology.

11.4 Neuroradiology

Ultrasound can be performed for neonates by using a high-frequency transducer through the anterior fontanelle in coronal and sagittal planes. The most common findings include an irregular and echogenic ependyma, the presence of intraventricular debris and stranding, intraventricular adhesions and septae associated with ventricular dilatation [58].

Noncontrast CT may be nonspecific which include dependent hyperdense ventricular debris, univentricular or biventricular hydrocephalus, periventricular low density and features of underlying abnormality. Contrast enhanced CT shows homogenous enhancement of the ependymal lining of the ventricles (**Figure 14**).

MRI findings are similar to CT, with the ventricular debris hyperintense T1 weighted images and hypointense on T2 weighted images, with high signal on DWI, reduced ADC value and are seen as fluid levels of high signal intensity on FLAIR images [57].

11.5 Treatment

On clinical suspicion of nosocomial ventriculitis, empiric antibiotic therapy should be initiated.

Surgical modalities for ventriculitis include a] intraventricular/intrathecal antibiotics b] continuous intraventricular irrigation therapy using a closed drainage system c] neuroendoscopic ventricular lavage with septostomy/monroplasty [59].

Rational for early surgery in ventriculitis: The rapid development of polydrug-resistant class of gram-negative bacteria, poor passage of antibiotics in the intraventricular space after intravenous injection and hampered CSF flow from ventricles along with infection cause the ventricles to become an enclosed system in

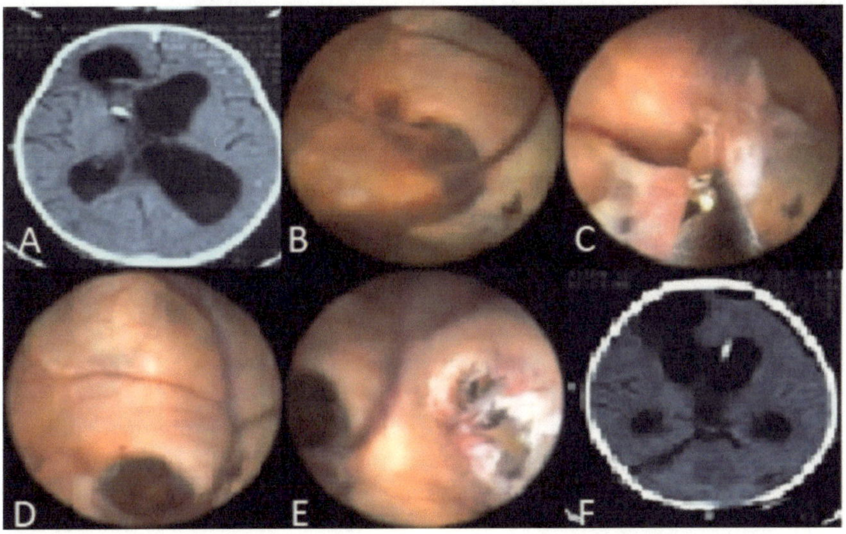

Figure 14
5 year female child presented with A] monoventricular [left lateral ventricle] hydrocephalus post EVD insertion. B] Intraoperative evidence of exudates causing blockage of left foramen of Monro. C] Intraventricular exudates were cleared. D] Foraminoplasty done with E] Septostomy. As there was evidence of exudates in the right lateral ventricle which could be seen from the septostomy, neuroendoscopic lavage was given bilaterally. EVD was inserted on the left side. F] Immediate postoperative radiology showing bilaterally equal ventricles with resolving hydrocephalus.

which infected CSF collects and the low concentration of antibiotics after intravenous injection is insufficient to kill the bacteria [60].

The endoscope allows seeing the entire cavity of the ventricles and also facilitating endoscopic lavage and clearance of the infected CSF. The decrease in infective load increases the effectiveness of the intrathecal antibiotics. Lavage inside the ventricles with continuous irrigation over a prolonged period has been reported with favorable outcomes.

Authors recommendations: We are of the opinion that in cases of ventriculitis a biventricular lavage with removal of ventricular exudates and adhesions is better than dealing the infection through one ventricle only.

Case Illustration (**Figure 14**).

12. Central neurocytoma

12.1 Introduction

Central neurocytoma is a rare benign tumor of neuronal differentiation that is classified as grade II by World Health Organization [WHO]. They comprise about 0.1 to 0.5% of all brain tumor and are typically seen in young adults around the third decade. They are characteristically located in the supratentorial ventricular system more commonly involving the foramen of Monro and in lesser frequency in the third and fourth ventricle [61].

12.2 Clinical features

These patients present with signs and symptoms of raised ICP induced by obstructive hydrocephalus [61, 62]. Typical clinical symptoms include headache, nausea, vomiting, seizures, paresthesias, balance problems, decreased consciousness, weakness, memory and visual disturbances. In rare cases, IVH may also occur.

12.3 Neuroradiology

On noncontrast CT, central neurocytoma appears hyperdense with calcifications seen in around 50% of cases which are usually punctate in nature [62, 63]. Cystic degeneration, seen usually in larger tumors can lead to heterogenous appearance of the tumor. CNs have mild to moderate enhancement on contrast enhanced CT. On MRI, CNs are hypo to isointense on T1, iso to hyperintense on T2, with moderate contrast enhancement (**Figure 15**).

12.4 Treatment

Surgical management with gross total resection is the treatment of choice [62, 64]. Goals of surgery are a] to establish the CSF pathway, determine the histopathological diagnosis and establish maximal surgical resection with minimum risk of neurological impairment. Radiotherapy to the tumor bed is debatable and is indicated in cases with incomplete resection to prevent tumor progression and recurrence. Stereotactic radiosurgery has also been proved to be effective compared to radiotherapy in terms of diminishing tumor recurrences and radiation associated complications in these cases.

In a study by Chun li et al. on 9 cases of central neurocytoma 2 patients died in the postoperative period. They concluded that the possibility of recurrence of

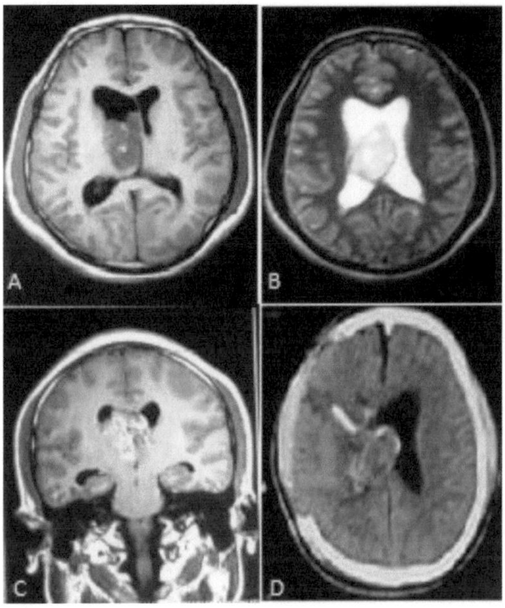

Figure 15.
30 year male patient presented with features of raised ICP with MRI findings suggestive of a lesion arising from the septum pellucidum causing obstruction at the level of right foramen of Monro. Endoscopic surgery was planned to decompress the tumor with biopsy. However owing to intraoperative intratumoral hemorrhage, a decompressive craniectomy had to be performed and the tumor was decompressed through transcortical route. Biopsy was suggestive of central neurocytoma [WHO grade 2] with MIB1-Labeling index following which patient was referred for radiotherapy.

central neurocytomas should be considered based on the histologic features, especially proliferation index [MIB1-LI] [65].

Case Illustration (**Figure 15**).

13. Functionally isolated ventricles

In some cases, rapid therapeutic drainage of one lateral ventricle particularly after a low pressure VP shunt, can cause an ipsilateral slit ventricle which functionally obstructs the foramen of Monro [66]. This functional obstruction may lead to dilatation of the contralateral ventricle.

The treatment of choice of functional ventriculomegaly is neuroendoscopic intervention with septostomy with third ventriculostomy (**Figure 16**).

Case Illustration (**Figure 16**).

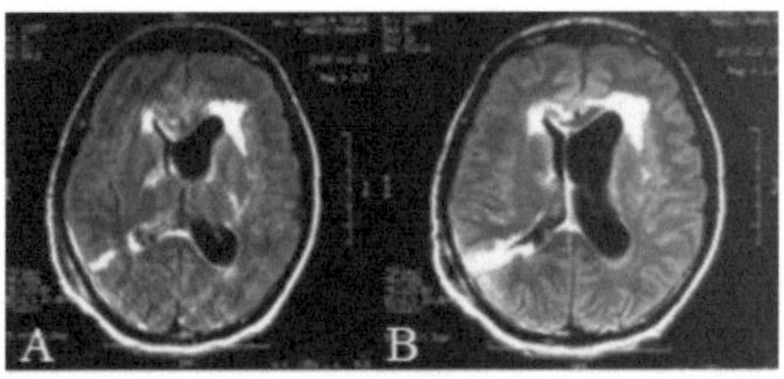

Figure 16.
34 year male, recently shunted for TBM with hydrocephalus was referred with altered sensorium. MRI brain was suggestive VP shunt in situ in right lateral ventricle and asymmetric dilatation of the left lateral ventricle. Neuroendoscopy did not reveal any obstruction at the foramen of monro and hence septostomy was done following which patient improved.

14. Subependymomas

14.1 Introduction

Subependymomas are rare, benign, indolent tumors of ependymal origin which comprise of 0.2%–0.7% of all intracranial tumors. They are slow growing neoplasms frequently seen in middle aged men more commonly in the fourth [50–60%] and the lateral [30–40%] ventricles. They also have a predilection for the spine and are seen in the cervical and cervicothoracic region [67, 68].

These tumors are frequently asymptomatic found incidentally on autopsy or on neuroimaging done for other medical indications.

14.2 Clinical features.

Symptomatic presentation depends on the location and size of the lesion. According to literature, tumors located at the septum pellucidum and the foramen of Monro with a size of more than 4 cms were more likely to become symptomatic with signs of raised ICP. Cases with spontaneous intratumoral hemorrhage have also been reported causing obstructive hydrocephalus. Despite the benign nature of the tumor, cases of tumor recurrence and CNS metastasis have also been reported [69, 70].

14.3 Neuroradiology

Radiological appearance of these tumors can vary depending on their location. On MRI, these tumors are typically well demarcated nodular lesions typically in the fourth or lateral ventricle which are hypo to isointense on T1 weighted images, hyperintense on T2 weighted images with minimum contrast enhancement [68, 69].

14.4 Treatment

Total surgical resection is the treatment of choice as these tumors are well demarcated, noninvasive and avascular. Role of postoperative radiation is debatable and is reserved in cases with subtotal surgical resection or in case of recurrence [71, 72].

15. Idiopathic Foramen of Monro stenosis

Idiopathic stenosis of foramen of Monro is a rare cause of enlargement of the lateral ventricles. It can occur due to an absent or stenosed foramen of Monro or when a membrane occludes a normal sized foramen of Monro. Patients are usually asymptomatic but can occasionally present with symptoms of raised ICP [71, 73].

Contrast enhanced MRI is the investigation of choice to exclude other causes of foramen of Monro obstruction and to assess the presence of potential membranes. In addition Cine-MRI CSF flowmetry is done to display CSF flow dynamics and velocity and to determine the site of CSF flow obstruction [73].

Neuroendoscopy is the treatment of choice as endosopic fenestration, foraminoplasty and septostomy can be done safely and effectively with this technique and spare the patient lifelong cumulative risk of shunt failure [74].

16. Choroid plexus tumors

16.1 Introduction

Choroid plexus tumors [CPTs] are rare intraventricular papillary tumors derived from choroid plexus epithelium which account for 0.3%–0.7% of all intracranial tumors, 2–6% of pediatric brain tumors, 10–20% of brain tumors in children less than 1 year and less than 1% of all adult intracranial tumors. The spectrum of choroid plexus tumors include a] WHO grade I choroid plexus papilloma b] WHO grade II atypical choroid plexus papilloma and c] WHO grade III choroid plexus carcinoma. Choroid plexus papillomas are more common than choroid plexus carcinomas in ratio of about 5: 1 [72, 75].

CPTs arise wherever choroid plexus tissue exists; the lateral ventricle being the most common site [50% of cases] followed by fourth ventricle [40%], third ventricle [5%] and multiple ventricles [5%] [75, 76].

16.2 Clinical features

CPTs most commonly present with signs of raised ICP and hydrocephalus [77–79]. Hydrocephalus can result due to overproduction of CSF, IVH, obstruction of CSF flow or impaired absorption. In addition to symptoms of raised intracranial pressure, patients may also present with seizures, focal neurological deficits, complications of chronic raised intracranial pressure like CSF rhinorrhea and visual disorientation, macrocephaly, and gait unsteadiness. Specific clinical symptoms can also be present according to the location of the tumor. Diencephalic seizures and bobble head doll syndrome- due to compression of the thalamus, signs of brainstem compression, cranial nerve palsies and cerebellar dysfunctions- due to posterior fossa CPTs, endocrine disturbances, precocious puberty, diabetes insipidus or diencephalic disorders- due to tumor in the third ventricle.

16.3 Neuroradiology

On Non contrast CT [NCCT], these tumors are seen as isodense to hyperdense intraventricular masses with intense contrast enhancement [72, 80].

MRI is the investigation of choice for these tumors. On MRI, CPTs appear as a large intraventricular lesion with irregular enhancing margins. They are isointense to slightly hyperintense on T1 weighted images, slightly hyperintense on T2 weighted images and show significant contrast enhancement.

16.4 Treatment

Maximum surgical resection followed by non-standardized use of adjuvant chemotherapy and radiotherapy is the treatment of choice for CPTs [75, 81].

17. Management of Foramen of Monro obstruction

17.1 Diagnosis of foramen of Monro obstruction

A detailed clinico-radiological examination is mandatory for diagnosis of foramen of Monro obstruction. Clinical manifestations and radiological features of various pathologies causing foramen of monro obstruction have already been discussed earlier in this chapter. Radiological features suggestive of foramen of monro obstruction include a] univentricular or biventricular hydrocephalus b] non-visualization of the third ventricle c] obvious pathology seen in the third or lateral ventricle in the vicinity of the foramen.

17.2 Management

Nonoperative management: nonoperative management of foramen of Monro obstruction can be contemplated for very few indications like in patients with a small IVH causing hydrocephalus or in cases of idiopathic foramen of Monro obstruction with no signs of raised ICP.

17.3 Surgical management

Surgical management of the foramen of Monro obstruction depends on the pathology causing the obstruction. Approach can be done either by a] open craniotomy b] neuroendoscopy.

17.4 Open Craniotomy approaches

There are many surgical approaches for approaching tumors of the ventricular system. The open approach to foramen of Monro lesions depends on whether the lesion is predominant in the lateral or third ventricle. Microneurosurgical techniques used to reach the frontal horn of the lateral ventricles and foramen of Monro are a] frontal transcortical approach and b] anterior transcallosal approach [82]. Approach to foramen of monro lesions predominantly in third ventricle can be done by lamina terminalis approach either through the pterional or subfrontal corridors (**Figure 17**).

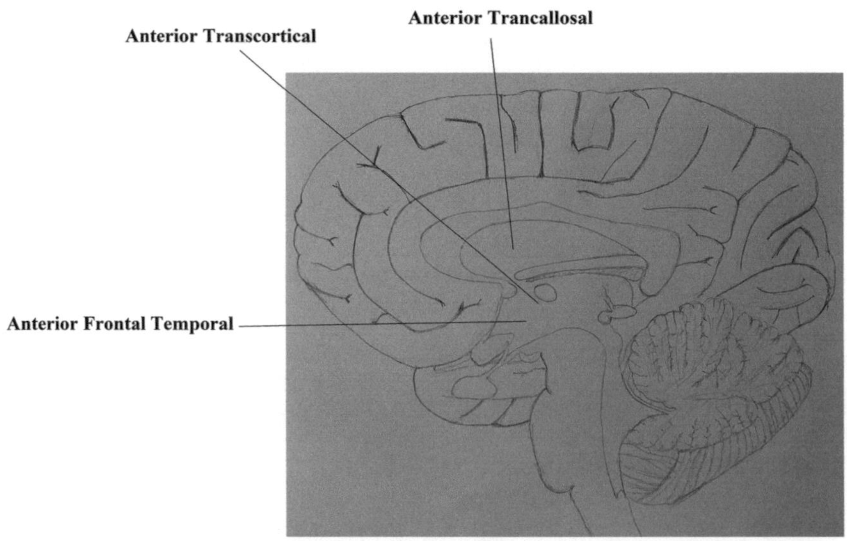

Figure 17.
Surgical approaches to the frontal horn of the lateral ventricle or third ventricle.

18. Endoscopic approach

18.1 Endoscopic anatomy

Anatomical landmarks important to be recognized while performing neuroendoscopic approach to the lateral and third ventricle for performing septostomy and foraminoplasty include: the anterior caudate vein, thalamostriate vein, septal vein, choroid plexus and foramen of Monro (**Figure 18**).

18.2 Introduction

With recent advances in endoscopic techniques, neuroendoscopy has become the first line of management for intraventricular pathologies. Endoscopic approach to the foramen of Monro can be considered as minimally invasive version of the open transcortical approach with the added advantage of panoramic view which can be achieved by angling the scope or using scopes of different viewing angles. With recent advances in scope instrumentation, bimanual dissection of pathologies has also become possible. The definitive treatment of the lesion by endoscopic approach can always be supplemented with septostomy and monroplasty/foraminoplasty.

18.3 Septostomy

Endoscopic septostomy allows to bypass a unilateral foramen of Monro obstruction creating a CSF circulation between the obstructed ventricle and the opposite lateral ventricle that communicates with the third ventricle by the normal foramen of Monro [77]. This communication between both the lateral ventricles converts them into a single compartment thus allowing both the ventricles to be drained by one shunt in cases of bilateral foramen of Monro obstruction.

For performing a septostomy, a linear incision is generally taken 5-6cms lateral to the midline which is more lateral than the incision taken for endoscopic third ventriculostomy. The use of navigation also helps in deciding the site of the incision. A semicircular incision may be opted if we are planning to insert an Ommaya reservoir.

Authors recommendations: The site of septostomy on the septum pellucidum is generally posterosuperior to the foramen of Monro, posterior to the anterior septal vein, at the point where the septum appears to be avascular and thinned out. The

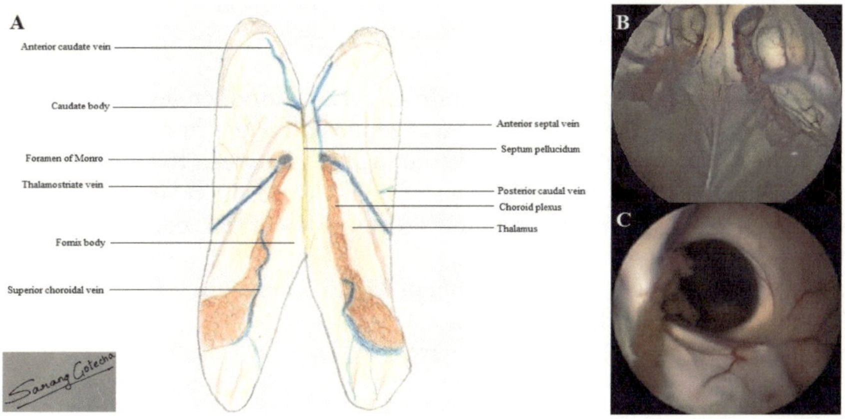

Figure 18.
Anatomy of the lateral ventricle.

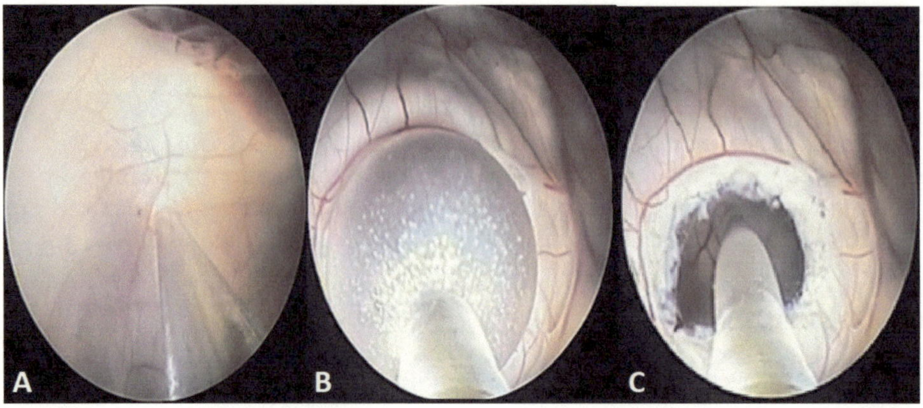

Figure 19.
A] Septum pelucidum being probed in a avascular area B] post- perforation, dilatation of septostomy with balloon dilator C] opposite ventricular wall seen through septostomy.

probing of the septum with the blunt tip of the bipolar probe also gives the surgeon an idea of the thickness of septum pellucidum. If tumor biopsy or resection is planned simultaneously, septostomy should precede the biopsy or resection as bleeding while dealing with the tumor may obscure the anatomical orientation making the septostomy difficult (**Figure 19**).

Intraventricular hemorrhage has been documented as a complication in literature and the possible reason is injury to the contralateral septal vein during the septostomy being performed. However, no such experience has been encountered by the authors.

In a study by Oertel et al. septostomy was performed 5–10 mm posterior to the interventricular foramen, in the middle of the corpus callosum [CC] and the fornix [83]. In a study by Hamada et al. perforation of the septum was done between the anterior and posterior septal vein [78]. In a study by Roth et al. perforation was done in the anterior septal area, at the level of the interventricular foramen, midway between the corpus callosum and the fornix [79].

The largest study on the procedure of endoscopic septostomy was done by P.R. Aldana and their inference was that septostomy is the most adequate surgery for unilateral obstruction at the interventricular foramen level [84].

19. Endoscopic Foraminoplasty

Endoscopic foraminoplasty at the interventricular foramen level is not very commonly reported in the literature. It was reported for the first time by Oi et al. [85]. This procedure establishes back the connection between the lateral and the third ventricle. It obviates the need of a shunt and avoids its related complications. For performing foraminoplasty, the incision and entry of the scope is similar to that of septostomy.

On entry to the ventricle, the foramen of Monro is recognized by choroid plexus and the thalamostriate and septal veins.

Endoscopic foraminoplasty can be performed by:

a. excision of the offending lesion

b. clearance of the exudates, adhesions or blood products

c. perforation of septations or membranes

d. enlargement with balloon catheter or dilating forceps

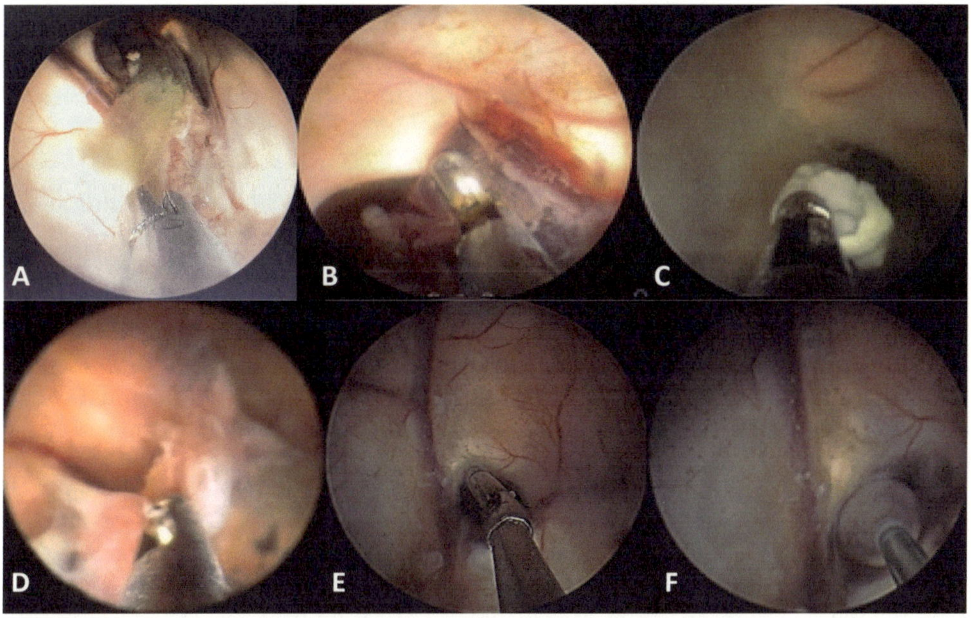

Figure 20.
foraminoplasty can be done by A] excision of lesion B] removal of blood products C] removal of exudates D] removal of adhesions E] dilatation with dilator forceps F] dilatation with balloon catheter.

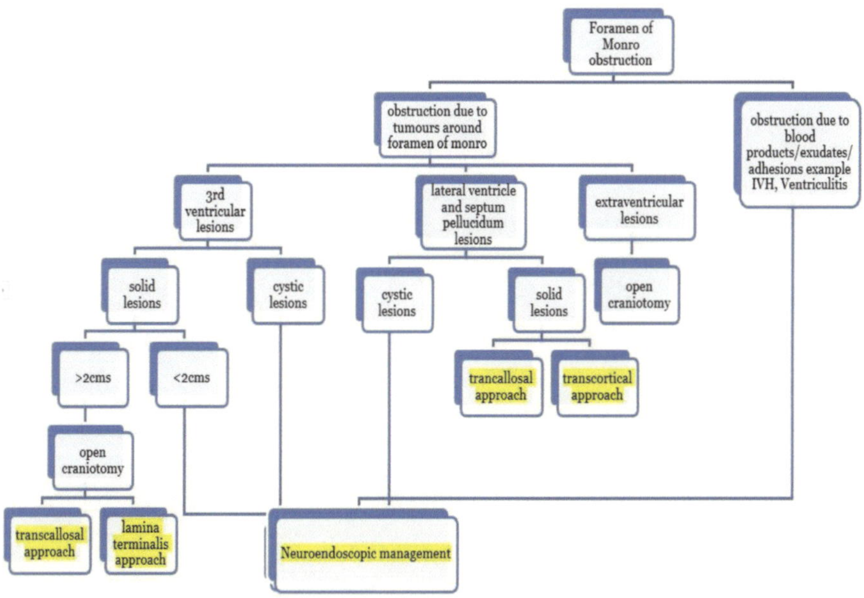

Table 4.
Algorithm for surgical management of foramen of Monro obstruction.

As it always carries a risk of recurrent obstruction because of the scarring, foraminoplasty should be supplemented with a septostomy. The procedure of dilatation carries the risk of injury to the septal vein or thalamostriate vein and there can also be injury to the fornix leading to memory problems. However, in our experience, we have never encountered vascular injury in any of our cases. In 1 case, injury to the fornix was seen. However, postoperatively, the patient did not experience any memory problems (**Figure 20**) (**Table 4**).

20. Conclusion

Lesions causing foramen of Monro obstruction resulting in hydrocephalus are a fairly uncommon entity encountered in neurosurgical practice with craniopharyngioma and colloid cysts being the most common pathology in children and adults respectively.

Treatment consists of open craniotomy for solid tumors and endoscopic approaches (transnasal transsphenoidal and cranial) for cystic lesions. Endoscopic approach is particularly helpful in decreasing the convalescence period and postoperative complications and thus, should be offered as a first line of treatment whenever suited.

Septostomy should be a part of standard treatment in all the patients having foramen of Monro obstruction so as to obviate the need of added shunt procedures.

Although small solid lesions less than 2 cms in size can be addressed by endoscopic approach, the learning curve required for endoscopic approach to deal with solid lesions is very steep. Thus, correct patient selection is of utmost importance for optimal patient outcome.

Author details

Ashish Chugh[1], Sarang Gotecha[1*], Prashant Punia[1] and Neelesh Kanaskar[2]

1 Department of Neurosurgery, Dr.D.Y.Patil Medical College, Pimpri, Pune, India

2 Department of Anatomy, Dr.D.Y.Patil Medical College, Pimpri, Pune, India

*Address all correspondence to: dr.sarangsgotecha@gmail.com

References

[1] Cai Q, Wang J, Wang L, Deng G, Chen Q, Chen Z. A classification of lesions around interventricular foramen and its clinical value. :8.

[2] Gray H, Williams PL. Gray's Anatomy: The Anatomical Basis of Medicine and Surgery. 38th ed. Churchill Livingstone, 1995; 1202–1205 p.

[3] Doctors Monro: A Medical Saga. By R. E. Wrightst. Clair, 8½ × 5½ in. Pp. 190, with 8 plates. 1964. London: Wellcome Historical Medical Library. 30s. British Journal of Surgery. 2005 Dec 8;52(4):317–317.

[4] Monro A. Observations on the structure and functions of the nervous system: illustrated with tables [Internet]. Edinburgh [u.a.]; 1783. X, 176 S., zahlr. Ill. Available from: https://doi.org/10.11588/diglit.4812

[5] Mokri B. The Monro-Kellie hypothesis: applications in CSF volume depletion. Neurology. 2001 Jun 26;56 (12):1746–8.

[6] Dutta AK. Essentials of Human Embryology. 5th ed. Current Books International.2006.; 261–263 p.

[7] Chugh A, Gotecha S, Punia P, Patil A, Kotecha M. Neuroendoscopic Excision of Third Ventricular Colloid Cysts. IJNNS. 2018;10(4):157–64.

[8] Ravnik J, Bunc G, Grcar A, Zunic M, Velnar T. Colloid cysts of the third ventricle exhibit various clinical presentation: a review of three cases. Bosn J of Basic Med Sci. 2014 Aug 14;14 (3):132.

[9] Tenny S, Thorell W. Colloid Brain Cyst. In: StatPearls [Internet]. Treasure Island (FL): StatPearls Publishing; 2021 [cited 2021 May 30]. Available from: http://www.ncbi.nlm.nih.gov/books/NBK470314/

[10] Armao D, Castillo M, Chen H, Kwock L. Colloid Cyst of the Third Ventricle: Imaging-pathologic Correlation. 2000;8.

[11] Decq P, Goutagny S, Staquet H, Iakovlev G, Krichen W, Faillot T, et al. Hydrocephalus and Colloid Cysts. In: Cinalli G, Özek MM, Sainte-Rose C, editors. Pediatric Hydrocephalus [Internet]. Cham: Springer International Publishing; 2019 [cited 2021 May 30]. p. 797–819. Available from: http://link.springer.com/10.1007/978-3-319-27250-4_13

[12] California Institute Of Behavioral Neurosciences & Psychology, Adjepong D. Interventricular Colliodal Cyst: A Literature Review. SCTI. 2020 May 20;4(2):1–4.

[13] Beaumont TL, Limbrick DD, Rich KM, Wippold FJ, Dacey RG. Natural history of colloid cysts of the third ventricle. JNS. 2016 Dec;125(6): 1420–30.

[14] Lewis AI, Crone KR, Taha J, van Loveren HR, Yeh H-S, Tew JM. Surgical resection of third ventricle colloid cysts: Preliminary results comparing transcallosal microsurgery with endoscopy. Journal of Neurosurgery. 1994 Aug;81(2):174–8.

[15] Levine N, Miller M, Crone K. Endoscopic Resection of Colloid Cysts: Indications, Technique, and Results during a 13-Year Period. Minim Invasive Neurosurg. 2007 Dec;50(6): 313–7.

[16] Nair S, Gopalakrishnan C, Menon G, Easwer H, Abraham M. Interhemispheric transcallosal transforaminal approach and its variants to colloid cyst of third ventricle: Technical issues based on a single institutional experience of 297 cases. Asian J Neurosurg. 2016;11(3):292.

[17] Fernandez-Miranda JC, Gardner PA, Snyderman CH, Devaney KO, Strojan P, Suárez C, et al. Craniopharyngioma: A pathologic, clinical, and surgical review. Eisele DW, editor. Head Neck. 2012 Jul; 34(7):1036–44.

[18] Lubuulwa J, Lei T. Pathological and Topographical Classification of Craniopharyngiomas: A Literature Review. J Neurol Surg Rep. 2016 Aug 22; 77(03):e121–7.

[19] Siomin V, Constantini S. Treatment of Hydrocephalus in Suprasellar Lesions. In: Cinalli G, Sainte-Rose C, Maixner WJ, editors. Pediatric Hydrocephalus [Internet]. Milano: Springer Milan; 2005 [cited 2021 Jun 14]. p. 163–70. Available from: https://doi.org/10.1007/978-88-470-2121-1_12

[20] Kassam AB, Gardner PA, Snyderman CH, Carrau RL, Mintz AH, Prevedello DM. Expanded endonasal approach, a fully endoscopic transnasal approach for the resection of midline suprasellar craniopharyngiomas: a new classification based on the infundibulum. JNS. 2008 Apr;108(4): 715–28.

[21] Müller HL. The Diagnosis and Treatment of Craniopharyngioma. Neuroendocrinology. 2020;110(9–10): 753–66.

[22] Deopujari CE, Karmarkar VS, Shah N, Vashu R, Patil R, Mohanty C, et al. Combined endoscopic approach in the management of suprasellar craniopharyngioma. Childs Nerv Syst. 2018 May;34(5):871–6.

[23] Ortiz Torres M, Shafiq I, Mesfin FB. Craniopharyngioma. In: StatPearls [Internet]. Treasure Island (FL): StatPearls Publishing; 2021 [cited 2021 Jun 14]. Available from: http://www.ncbi.nlm.nih.gov/books/NBK459371/

[24] Varlotto J, DiMaio C, Grassberger C, Tangel M, Mackley H, Pavelic M, et al. Multi-modality management of craniopharyngioma: a review of various treatments and their outcomes. Neuro-Oncology Practice. 2016 Sep 1;3(3): 173–87.

[25] Sterman H, Furlan AB, Matushita H, Teixeira MJ. Subependymal giant cell astrocytoma associated with tuberous sclerosis presenting with intratumoral bleeding. Case report and review of literature. Childs Nerv Syst. 2013 Feb;29 (2):335–9.

[26] Stavrinou P, Spiliotopoulos A, Patsalas I, Balogiannis I, Karkavelas G, Polyzoidis K, et al. Subependymal giant cell astrocytoma with intratumoral hemorrhage in the absence of tuberous sclerosis. J Clin Neurosci. 2008 Jun;15 (6):704–6.

[27] Moavero R, Pinci M, Bombardieri R, Curatolo P. The management of subependymal giant cell tumors in tuberous sclerosis: a clinician's perspective. Childs Nerv Syst. 2011 Aug; 27(8):1203–10.

[28] Moncef B. Management of subependymal giant cell tumors in tuberous sclerosis complex: the neurosurgeon's perspective. World J Pediatr. 2010 May;6(2):103–10.

[29] Kim J-Y, Jung T-Y, Lee K-H, Kim S-K. Subependymal Giant Cell Astrocytoma Presenting with Tumoral Bleeding: A Case Report. Brain Tumor Res Treat. 2017 Apr;5(1):37–41.

[30] Beaumont TL, Limbrick DD, Smyth MD. Advances in the management of subependymal giant cell astrocytoma. Childs Nerv Syst. 2012 Jul; 28(7):963–8.

[31] Shah H, Jain K, Shah J. Endoscopic excision of intraventricular neurocysticercosis blocking foramen of Monro bilaterally. Asian J Neurosurg. 2016;11(2):176.

[32] Arshad F, Rao S, Kenchaiah R, Prasad C, Shashidhar A. Intraventricular neurocysticercosis presenting as Bruns' syndrome: An uncommon presentation. Egypt J Neurol Psychiatry Neurosurg. 2020 Dec;56(1):54.

[33] Jensen TO, Post JJ. Intraventricular neurocysticercosis: Presentation, diagnosis and management. Asian Pacific Journal of Tropical Medicine. 2016 Aug;9(8):815–8.

[34] Shahani L, Mejia R, Garnes ND. Intraventricular Taenia solium Cysts Presenting with Bruns Syndrome and Indications for Emergent Neurosurgery. The American Journal of Tropical Medicine and Hygiene. 2015 Jun 3;92 (6):1261–4.

[35] Cuetter AC, Andrews RJ. Intraventricular neurocysticercosis: 18 consecutive patients and review of the literature. FOC. 2002 Jun;12(6):1–7.

[36] Rodriguez S, Dorny P, Tsang VCW, Pretell EJ, Brandt J, Lescano AG, et al. Detection of Taenia solium Antigens and Anti–T. solium Antibodies in Paired Serum and Cerebrospinal Fluid Samples from Patients with Intraparenchymal or Extraparenchymal Neurocysticercosis. J Infect Dis. 2009 May 1;199(9):1345–52.

[37] Psarros TG, Krumerman J, Coimbra C. Endoscopic management of supratentorial ventricular neurocysticercosis: case series and review of the literature. Minim Invasive Neurosurg. 2003 Dec;46(6):331–4.

[38] Chan KH, Cheung RTF, Fong CY, Tsang KL, Mak W, Ho SL. Clinical relevance of hydrocephalus as a presenting feature of tuberculous meningitis. QJM. 2003 Sep 1;96(9): 643–8.

[39] Raut T, Garg RK, Jain A, Verma R, Singh MK, Malhotra HS, et al. Hydrocephalus in tuberculous meningitis: Incidence, its predictive factors and impact on the prognosis. Journal of Infection. 2013 Apr;66(4): 330–7.

[40] Rajshekhar V. Management of hydrocephalus in patients with tuberculous meningitis. Neurol India. 2009;57(4):368.

[41] Sharma C, Acharya M, Kumawat BL, Kochar A. "Trapped temporal horn" of lateral ventricle in tuberculous meningitis. Case Reports. 2014 Apr 4;2014(apr03 2): bcr2014203837–bcr2014203837.

[42] Goel A. Tuberculous meningitis and hydrocephalus. :1.

[43] Chugh A, Husain M, Gupta RK, Ojha BK, Chandra A, Rastogi M. Surgical outcome of tuberculous meningitis hydrocephalus treated by endoscopic third ventriculostomy: prognostic factors and postoperative neuroimaging for functional assessment of ventriculostomy. J Neurosurg Pediatr. 2009 May;3(5):371–7.

[44] Rawska A, Sałek M, Nowakowska M, Bąk M, Jamroz-Wiśniewska A, Rejdak K. Successful surgical treatment in a patient with a giant pituitary macroadenoma accompanied by obstructive hydrocephalus. J Pre Clin Clin Res. 2019 Sep 27;13(3):130–3.

[45] Zhang D, Chen J, Li Z, Wang J, Han K, Hou L. Clinical features and management of nonfunctioning giant pituitary adenomas causing hydrocephalus. Oncotarget. 2018 Mar 16;9(20):15409–17.

[46] Verhelst J, Berwaerts J, Abs R, Dua G, Weyngaert DVD, Mahler Ch. Obstructive Hydrocephalus as Complication of a Giant Nonfunctioning Pituitary Adenoma: Therapeutical Approach. Acta Clinica Belgica. 1998 Jan;53(1):47–52.

[47] Iglesias P, Rodríguez Berrocal V, Díez JJ. Giant pituitary adenoma: histological types, clinical features and therapeutic approaches. Endocrine. 2018 Sep;61(3):407–21.

[48] Bashari WA, Senanayake R, Fernández-Pombo A, Gillett D, Koulouri O, Powlson AS, et al. Modern imaging of pituitary adenomas. Best Practice & Research Clinical Endocrinology & Metabolism. 2019 Apr;33(2):101278.

[49] Ojha BK, Husain M, Rastogi M, Chandra A, Chugh A, Husain N. Combined trans-sphenoidal and simultaneous trans-ventricular-endoscopic decompression of a giant pituitary adenoma: case report. Acta Neurochir. 2009 Jul;151(7):843–7.

[50] Hanley D, Naff N, Harris D. Intraventricular Hemorrhage: Presentation and Management Options. Seminars in Cerebrovascular Diseases and Stroke. 2005 Sep;5(3):209–16.

[51] Bu Y, Chen M, Gao T, Wang X, Li X, Gao F. Mechanisms of hydrocephalus after intraventricular haemorrhage in adults. Stroke Vasc Neurol. 2016 Mar;1(1):23–7.

[52] Sahni R, Weinberger J. Management of intracerebral hemorrhage. Vasc Health Risk Manag. 2007;3(5):701–9.

[53] Nyquist P. Management of acute intracranial and intraventricular hemorrhage. Crit Care Med. 2010 Mar;38(3):946–53.

[54] Basaldella L, Marton E, Fiorindi A, Scarpa B, Badreddine H, Longatti P. External ventricular drainage alone versus endoscopic surgery for severe intraventricular hemorrhage: a comparative retrospective analysis on outcome and shunt dependency. Neurosurg Focus. 2012 Apr;32(4):E4.

[55] Beer R, Lackner P, Pfausler B, Schmutzhard E. Nosocomial ventriculitis and meningitis in neurocritical care patients. J Neurol. 2008 Nov;255(11):1617–24.

[56] Humphreys H, Jenks PJ. Surveillance and management of ventriculitis following neurosurgery. Journal of Hospital Infection. 2015 Apr;89(4):281–6.

[57] Harris L, Munakomi S. Ventriculitis. In: StatPearls [Internet]. Treasure Island (FL): StatPearls Publishing; 2021 [cited 2021 May 30]. Available from: http://www.ncbi.nlm.nih.gov/books/NBK544332/

[58] Mohan S, Jain KK, Arabi M, Shah GV. Imaging of Meningitis and Ventriculitis. Neuroimaging Clinics of North America. 2012 Nov;22(4):557–83.

[59] Tabuchi S, Kadowaki M. Neuroendoscopic surgery for ventriculitis and hydrocephalus after shunt infection and malfunction: Preliminary report of a new strategy. Asian J Endosc Surg. 2015 May;8(2):180–4.

[60] Remeš F, Tomáš R, Jindrák V, Vaniš V, Šetlík M. Intraventricular and lumbar intrathecal administration of antibiotics in postneurosurgical patients with meningitis and/or ventriculitis in a serious clinical state: Clinical article. JNS. 2013 Dec;119(6):1596–602.

[61] Runderawala H, Kantharia A, Oak P, Mahore A. Central Neurocytoma - A Rare Brain Tumor. J Assoc Physicians India. 2018 Apr;66(4):77–8.

[62] Lee SJ, Bui TT, Chen CHJ, Lagman C, Chung LK, Sidhu S, et al. Central Neurocytoma: A Review of Clinical Management and Histopathologic Features. Brain Tumor Res Treat. 2016;4(2):49.

[63] Ravikanth R. Neuroradiological and histopathological findings of intraventricular central neurocytoma. CHRISMED J Health Res. 2017;4(2):125.

[64] Marri M, Ahmad I, Ahmad K, Ashfaq Z. Central neurocytoma of the third ventricle: Case report and treatment review. Egyptian Journal of Basic and Applied Sciences. 2017 Dec;4 (4):361–5.

[65] Chen C-L, Shen C-C, Wang J, Lu C-H, Lee H-T. Central neurocytoma: a clinical, radiological and pathological study of nine cases. Clin Neurol Neurosurg. 2008 Feb;110(2):129–36.

[66] Atalay B, Yilmaz C, Cekinmez M, Altinors N, Caner H. Treatment of hydrocephalus with functionally isolated ventricles. Acta Neurochir (Wien). 2006 Dec;148(12):1293–6.

[67] Jain A, Amin AG, Jain P, Burger P, Jallo GI, Lim M, et al. Subependymoma: clinical features and surgical outcomes. Neurological Research. 2012 Sep;34(7): 677–84.

[68] Hernandez-Duran S, Yeh-Hsieh T-Y, Salazar-Araya C. Pedunculated intraventricular subependymoma: Review of the literature and illustration of classical presentation through a clinical case. Surg Neurol Int. 2014;5(1):117.

[69] Varma A, Giraldi D, Mills S, Brodbelt AR, Jenkinson MD. Surgical management and long-term outcome of intracranial subependymoma. Acta Neurochir. 2018 Sep;160(9):1793–9.

[70] Schroeder J, LeFever D, Entezami P, Mrak RE. Multiple supratentorial subependymomas causing obstructive hydrocephalus. BMJ Case Reports. 2017 Jun 3;bcr-2016-215625.

[71] Gomez-Ruiz N, Polidura MC, Crespo Rodriguez AM, Arrazola García J. Idiopathic stenosis of foramina of Monro in an asymptomatic adult patient: a rare entity radiologists should be aware of. BJR|case reports. 2020 Jun; 6(2):20190102.

[72] Jaiswal S, Behari S, Jain V, Vij M, Mehrotra A, Kumar B, et al. Choroid plexus tumors: A clinico-pathological and neuro-radiological study of 23 cases. Asian J Neurosurg. 2013;8(1):29.

[73] Boruah DK, Arora M, Prakash A, Baishya H, Chakraborty P. IDIOPATHIC UNILATERAL FORAMEN OF MONRO STENOSIS: NEUROIMAGING FINDINGS IN THREE PATIENTS. jebmh. 2016 Apr 28;3(34):1673–5.

[74] Kalhorn SP, Strom RG, Harter DH. Idiopathic bilateral stenosis of the foramina of Monro treated using endoscopic foraminoplasty and septostomy. FOC. 2011 Apr;30(4):E5.

[75] Cannon DM, Mohindra P, Gondi V, Kruser TJ, Kozak KR. Choroid plexus tumor epidemiology and outcomes: implications for surgical and radiotherapeutic management. J Neurooncol. 2015 Jan;121(1):151–7.

[76] Lam S, Lin Y, Cherian J, Qadri U, Harris DA, Melkonian S, et al. Choroid Plexus Tumors in Children: A Population-Based Study. Pediatr Neurosurg. 2013;49(6):331–8.

[77] Giammattei L, Aureli V, Daniel R-T, Messerer M. Neuroendoscopic septostomy: Indications and surgical technique. Neurochirurgie. 2018 Jun;64 (3):190–3.

[78] Hamada H, Hayashi N, Kurimoto M, Umemura K, Hirashima Y, Endo S. Neuroendoscopic septostomy for isolated lateral ventricle. Neurol Med Chir (Tokyo). 2003 Dec;43(12):582–7; discussion 588.

[79] Roth J, Olasunkanmi A, Rubinson K, Wisoff JH. Septal vein symmetry: implications for endoscopic septum pellucidotomy. Neurosurgery. 2010 Dec;67(2 Suppl Operative):395–401.

[80] Sun MZ, Oh MC, Ivan ME, Kaur G, Safaee M, Kim JM, et al. Current management of choroid plexus carcinomas. Neurosurg Rev. 2014 Apr; 37(2):179–92.

[81] Menon G, Nair S, Baldawa S, Rao R, Krishnakumar K, Gopalakrishnan C. Choroid plexus tumors: An institutional series of 25 patients. Neurol India. 2010; 58(3):429.

[82] Tubbs RS, Oakes P, Maran IS, Salib C, Loukas M. The foramen of Monro: a review of its anatomy, history, pathology, and surgery. Childs Nerv Syst. 2014 Oct;30(10):1645–9.

[83] Oertel JMK, Schroeder HWS, Gaab MR. Endoscopic stomy of the septum pellucidum: indications, technique, and results. Neurosurgery. 2009 Mar;64(3): 482–91; discussion 491-493.

[84] Aldana PR, Kestle JRW, Brockmeyer DL, Walker ML. Results of endoscopic septal fenestration in the treatment of isolated ventricular hydrocephalus. Pediatr Neurosurg. 2003 Jun;38(6):286–94.

[85] Oi S, Enchev Y. Neuroendoscopic foraminal plasty of foramen of Monro. Child's nervous system : ChNS : official journal of the International Society for Pediatric Neurosurgery. 2008 Sep 1;24: 933–42.

Section 4

Infection

Chapter 6

Infections in CSF Shunts and External Ventricular Drainage

Roger Bayston

Abstract

Infection in those with hydrocephalus shunts or external drains (EVDs) can cause serious central nervous system damage with lasting sequelae. The infections usually involve bacterial colonisation and biofilm formation in the catheters. The nature and sources of pathogens and preventive measures are discussed. The risks of infection in shunts and EVDs is different. Infection in shunts is almost always initiated at their insertion or revision (exceptions are described). In contrast, in EVDs, the risk of infection persists throughout their use. The pathogen profile is also different. These factors are important considerations when planning preventive measures. Newer strategies such as antimicrobial catheters are discussed. Diagnosis of EVD infections in an already ill patient is difficult but guidelines can be useful. Treatment of the shunt and EVD infections are also addressed, with reference to modes and routes of antibiotic administration.

Keywords: Hydrocephalus, shunt, external ventricular drain, infection, biofilm, diagnosis, treatment, prevention, prophylactic antibiotics, antimicrobial catheters

1. Introduction

Though several historical attempts had been made to drain excess cerebrospinal fluid (CSF) in cases of hydrocephalus, this remained largely unsuccessful until the advent of valved shunting devices in the 1950's in USA. More recently endoscopic third ventriculostomy has been used in selected patients, but shunting is still the usual method of treatment of hydrocephalus. In patients with raised intracranial pressure due to trauma, malignancy or haemorrhage, where it is hoped the situation is temporary, external ventricular drainage (EVD) is often used. This temporary method of control of intracranial pressure is also used after shunt removal for infection, before insertion of a new shunt. The risks for infection in the two modes of treatment are different.

Infection in shunts appeared soon after they became more widely used [1], though for some time their cause and treatment remained poorly understood, until it was realised in the early 1970's that most were caused by a bacterium, *Staphylococcus epidermidis*, that hitherto had been considered a harmless commensal and common culture contaminant [2]. The mechanisms of infection and reasons for difficulty in treatment have been clarified over the subsequent decades.

EVD has a very long history, but infection remained a major problem until the introduction of sterile closed systems of drainage in 1941 [3]. It is still a matter of

IntechOpen

concern and more recent increases in infections due to multi-drug-resistant (MDR) bacteria have exacerbated this.

Infections in shunts lead to repeated operations and courses of antibiotics and can lead to further cognitive impairment. Infections in EVDs lead to longer hospital stay, courses of antibiotics, and worse overall neurosurgical outcomes. In both cases death can result. Prompt diagnosis and appropriate treatment are essential, and prevention should be the primary goal. These can be achieved best with an understanding of the underlying science.

2. Aetiology and incidence

Infection rates have fallen in both shunting and EVD since the 1970's when up to 23% of shunts were reported as becoming infected [4, 5]. More recent rates for shunt infection have been below 10%, with 6% reported in a clinical trial [6]. However, it has been clear for some time that the infection rate in infants shunted when less than 6 months of age is significantly higher [7, 8] sometimes approaching 15–20% of operations [9].

The reported infection rate in EVDs is very variable, mainly due to difficulties in diagnosis, diagnostic criteria used and significant differences in underlying pathologies between patient groups studied. Earlier studies reported higher rates, 15–23% [10, 11] while slightly later studies reported 7.5% more in keeping with our own observations [12, 13].

2.1 Causative organisms in shunt infection

Since the first reports of shunt infection, staphylococci have predominated in shunt infection, with the majority being coagulase – negative staphylococci (CoNS). Of these, most are *Staphylococcus epidermidis*. A minority of staphylococci are *Staphylococcus aureus*. The proportion of these that are methicillin- resistant (MRSA) varies between countries according to national MRSA epidemiology [14], but in most countries especially The Netherlands, Scandinavia and United Kingdom, the proportion of MRSA in shunt infections is low [15]. However, methicillin resistance in CoNS is now common [16, 17]. Another important shunt pathogen is *Cutibacterium (Propionibacterium) acnes* [18, 19], found mainly in adolescents and adults. This bacterium is under-reported and probably accounts for some of the "culture-negative" shunt infections, as it is anaerobic and slow-growing, taking up to 14 days to appear in culture. Infection with gram negative bacteria such as *Escherichia coli* and *Klebsiella pneumoniae* is less common [15, 16, 20] and probably occurs more commonly in shunted infants than in adults [21].

2.2 Causative organisms in EVD infection

Most cases of ventriculitis associated with EVD use are caused by staphylococci [13, 22, 23] but there is evidence that the proportion of gram negative bacteria might be increasing. Chatzi et al. [24] reported 81% of their EVD infections were due to MDR gram negative bacteria, mainly *Acinetobacter baumannii*; similar proportions were reported by others [25, 26]. Another matter of concern is the increase in enterococcal infections, reflecting the rise of this MDR gram positive bacterium in general surgical site infections. Notable differences between pathogens in shunts and EVDs are the increasing proportion of MDR gram negative bacteria and the occurrence of polymicrobial infections in EVD, uncommon in shunts.

3. Mechanisms of infection

3.1 Shunts

The main source of pathogens in shunts is the patient's skin [8, 27]. Skin commensals such as CoNS and *C acnes* cannot be eradicated by skin preparation, and they easily enter the incision where they are able to gain access to the shunt and possibly the ventricular system during shunt insertion. Once inside the shunt tubing, they attach to the surface of the silicone, after which they begin to proliferate. This proliferation is slow, because the carbon and nitrogen sources in CSF are insufficient to support vigorous bacterial growth, and in particular it has a very low iron content [28]. However, the plaques of bacteria eventually develop into a biofilm. Biofilms are communities of micro-organisms usually attached to a surface, and they are very common in device infections and in the environment. They are also the preferred mode of growth, rather than the very artificial growth conditions applied in the laboratory. It is interesting that the first report of a biofilm in a medical device was from a shunt infection [29]. This early report was produced in response to the need to explain why antibiotic treatment, shown to be effective against shunt pathogens in the laboratory, was ineffective in treating shunt infections. It is now generally accepted that biofilms explain this difficulty, which is found in infections in other implants. The early report postulated that the biofilm structure was maintained by a glycosaminoglycan produced by the bacteria, and that antibiotics were unable to penetrate this effectively. Further research has confirmed the chemistry of the biofilm matrix (though it is now accepted that other components are present). Later studies have confirmed the presence of bacterial biofilms in infected shunts [30, 31]. It is now realised that most antibiotics can penetrate bacterial biofilms effectively [32], but that they fail to kill the constituent bacteria [33–35]. This is because, when bacteria attach to a surface and develop a biofilm, they change their metabolism in order to conserve energy, and this involves downregulating all inessential functions such as cell wall synthesis, most protein synthesis and DNA replication. All these are target sites for common antibiotics, and the concentration of antibiotic needed to even reduce the numbers of biofilm bacteria is 500 to 1000 fold higher than that found in the laboratory [32]. This explains why antibiotic treatment alone is usually ineffective against biofilm infections.

Bacteria are shed from biofilms and this is one way in which they might reach the ventricular system, but bacteria also spread along the inside surfaces of the tubing, and they might also be introduced from the incision during shunt insertion.

3.2 EVD

EVD infections also involve biofilm formation inside the tubing as well as externally to it in the subcutaneous tunnel, and all shunt and EVD pathogens including *C acnes* and *Acinetobacter baumannii* produce biofilms [36, 37].

3.3 Periods of risk for infection

Generally shunts are at risk of infection only at insertion or revision, but exceptions are postoperative CSF leak from the incision, and later skin erosion over the shunt or perforation of abdominal viscus. Skin erosion might be due to pressure in a debilitated patient, or to poor tissue coverage and skin health in premature infants, or to malnutrition. The viscus most often perforated by the lower catheter is the intestine. Reports in the literature often concern children and surprisingly, many are brought to the emergency room by parents worried that they have a parasitic infection, based on the lower catheter protruding from the anus [38]. Unlikely though this may seem, we have also seen two similar cases. There are often several bacteria of enteric origin, including anaerobes, in the CSF but the patients are often not as ill as might be expected.

It is generally agreed that the risk to shunts from bacteraemia during dental treatment is extremely small and does not warrant antibiotic prophylaxis. Haematogenous infection in both VP and VA shunts is very uncommon.

The period of risk for EVDs is very different. Access by skin bacteria is possible during insertion of the EVD, but the main risk extends for the time the EVD is in place, and is from skin bacteria that migrate from the exit site, from interventions such as CSF sampling and drug administration, flushing for blockage and changes of collection system.

4. Diagnosis

4.1 Shunt infections

The features of infection are different in VP and VA shunts. The discharge of bacteria and inflammatory products from an infected VP shunt into the peritoneal cavity triggers a local inflammatory response that often results in obstruction of the outflow of CSF. Sometimes this involves the greater omentum, and a CSF-filled cyst is formed around the end of the shunt [39]. This inflammatory response causes distal-end blockage of the shunt and return of the symptoms of hydrocephalus. This is the main reason for the important difference in time of presentation between infected VP and VA shunts: VP shunt infections usually present within months of operation, while VA shunt infections can present years later. However, CSF pseudocysts can present many years after shunt insertion with no evidence of infection [40].

In view of the presenting symptoms in VP shunt infection being those of raised intracranial pressure, it is important to distinguish between a non-infected blockage and one arising from shunt infection. Features of VP shunt infection include fever, headache, vomiting and irritability, though all of these are variable in consistency and can be due to non-infective obstruction. If the symptoms appear within 6–8 months of insertion or revision, this increases the likelihood of infection. If there is erythema over the catheter track then this is an important sign, but it is not always present. Abdominal ultrasound may show adhesions or cyst formation. Shunt aspiration will reveal bacteria on gram film and/or culture, but will not always show raised white cell count or other CSF abnormalities. In the absence of clear indications, there is often reluctance to aspirate the shunt due to concern for introduction of infection, but this risk is slight. Blood culture is usually negative. Blood C-reactive protein (CRP) is useful as it is often raised as part of the tissue inflammatory response. The features of VP shunt infection therefore include:

Presentation <6-8 months of operation
Symptoms of shunt obstruction
Erythema over the catheter track
Raised C-reactive protein
Pyrexia
Bacteria in gram stain and culture of aspirated CSF

Gram film examination is useful even if the CSF appears to be clear to the naked eye, as if bacteria can be seen it can make an early diagnosis irrespective of culture results. If culture is negative in the presence of a positive gram film then further measures can be taken such as extended anaerobic culture.

As not all VP shunt infections are contracted at operation, there will be some resulting from skin erosion over the valve or perforation of abdominal viscus but these are uncommon and the diagnosis is usually obvious. Haematogenous shunt infections are extremely rare. "Late" infections sometimes occur, due especially to *C acnes*, but those due to *Streptococcus pneumoniae*, *Neisseria meningitidis* or *Haemophilus influenzae* are almost always community-acquired meningitis in a person with a shunt, not a shunt infection, and this has important implications for treatment.

Presentation of VA shunt infection is often also within a few months of operation but in this case it can extend to several years later, leading to the unfounded suspicion that these infections are not contracted at surgery. Those VA shunt infections that present after more than 1–2 years are sometimes associated with immune complex disease. Here, the bacterial antigen discharging from the shunt into the bloodstream provokes an antibody response, and eventually the concentrations of circulating antigen and antibody combine into insoluble complexes that are deposited mainly on basement membranes [41, 42]. The skin, lungs, joints and renal glomerulae are particularly affected [43]. VA shunt infections can therefore present as chronic skin lesions (some haemorrhagic), chronic non-productive cough, swollen painful joints or haematuria. This often leads to initial referral to dermatology, respiratory medicine, rheumatology [44], orthopaedics and nephrology [45], and the diagnosis of shunt infection is sometimes missed or delayed. The causative bacteria in such cases are usually either CoNS or *C acnes*. Immune complex disease usually resolves on shunt removal.

Again aspiration of CSF from the shunt reservoir usually reveals the infecting bacterium. Blood cultures are usually positive but in very longstanding cases the pathogen might be non-culturable, or might appear as the biofilm phenotype known as small colony variants (SCV) which can be difficult to identify in the clinical laboratory [46, 47]. Iron-unresponsive anaemia is often a feature, and complement C3 and C4 levels are usually low due to complement consumption. CRP is often normal. An antibody assay has been used to diagnose late-presenting VA shunt infection [48, 49].

4.2 EVD infection

Diagnosis of infection in EVDs is often difficult. Features consistent with a diagnosis of ventriculitis are often present in patients with traumatic brain injury or stroke, and fever, with raised CSF white cell count, raised CSF protein level, disturbance of consciousness, Glasgow Coma Score, and inflammatory markers such as CRP are not helpful [50, 51]. In order to overcome the problem of raised white blood cell counts in patients with blood in the CSF, an index has been proposed, based on comparison of white cells and red cells in CSF and blood [52] but the number of patients in their study was small. A raised level of soluble Triggering Receptor Expressed on Myeloid cells (s-TREM) in CSF has been reported to be a reliable marker of ventriculitis even in the presence of haemorrhage [53] and this merits further investigation.

However, a positive culture result from CSF has been held to be the "gold standard," yet this is fraught with problems. Many isolates are skin commensals, and might be either pathogens or contaminants, or might be colonising the distal parts of the ventricular catheter but absent from the ventricles. If a recognised pathogen such as *S aureus* or *A baumannii* is isolated then this is generally taken as a reason to begin definitive treatment, but more than a single isolate of the same strain of CoNS is usually required. Isolates from broth cultures alone should usually be disregarded as likely contaminants, and broth cultures are generally unhelpful. Some have advocated daily CSF aspiration and examination [54] but this has not been found to be reliable in diagnosing or predicting ventriculitis [55], and has been identified as a risk factor for EVD infection. Therefore, in addition to a positive CSF culture, the Infectious Diseases Society of America (IDSA) recommendation is: "*New* headache, fever, evidence of meningeal irritation, seizures, and/or *worsening* mental status are suggestive of ventriculitis or meningitis in the setting of recent trauma or neurosurgery (strong, moderate evidence)" [56]. This guidance applies the need for the feature to be "new" and this is an important consideration in those patients already showing non-specific symptoms due to their underlying pathology.

5. Treatment

5.1 Shunt infections

There are several obstacles in the way of successful treatment of shunt infections. Though in most institutions, MRSA is not a common shunt pathogen, many CoNS are multi-resistant including to methicillin and other beta-lactam antibiotics. Another problem is the poor CSF penetration of most antibiotics given intravenously [57] so CSF levels are below the minimum bactericidal concentration (MBC). A third problem is the presence of shunt pathogens as biofilms in the catheter, as eradication of these requires up to 1000 times more antibiotic than the MBC measured in the laboratory [32]. The best chance of successful treatment of shunt infection is therefore shunt removal followed by a course of antibiotics and usually EVD before replacement with a new shunt if required [58]. This topic has been further confirmed by a review by James et al. [59]. A study in children showed 88% first-time cure using shunt removal, antibiotics and EVD, a lower success rate with an immediate shunt replacement protocol, but only 33% success with antibiotics and no shunt removal [60]. In many such studies there is also a disturbingly high mortality rate in those managed with shunt retention. Once the shunt is removed, this leaves a residue of infection in the ventricular system, and as it has arisen from the biofilm in the catheters it is likely to exhibit the biofilm phenotype and have a raised MBC. However, the problem of CSF penetration now becomes paramount. Ventriculitis caused by CoNS or *C acnes* does not give rise to vigorous inflammatory response [61], and this limits access of antibiotics to the CSF. Several factors apart from inflammation also influence the penetration of antibiotics into CSF [57]. Many antibiotics that could be used to treat ventriculitis fail to achieve sufficient concentrations in the CSF [62–64], and this has led to consideration of additional intraventricular administration via EVD [65]. Such a protocol was recommended by the British Society for Antimicrobial Chemotherapy [66] for staphylococcal infections, consisting of intraventricular vancomycin 20 mg daily, and oral or intravenous (IV) rifampicin 300 mg twice daily (15 mg/kg daily for children). The protocol has been shown to reduce the risk of relapse and to shorten the course of treatment needed [67, 68]. However, the association of variable to low antibiotic penetration with clinical failure has been questioned [69], but much of the information on antibiotic penetration into the CSF comes from patients with meningitis, and it is accepted that in ventriculitis the penetration is lower, especially when staphylococci are involved. A general principle is that if CSF antibiotic levels that reach the MIC can be achieved by intravenous administration, then this is sufficient, but the MBC is probably more important and this needs to be 5–8 times the MIC to ensure clinical success [70]. It is generally agreed that intraventricular administration of vancomycin is safe, irrespective of CSF levels, which can reach 100 mg/L, though some have advocated monitoring of CSF levels and dosage adjustment [68], which we have not found necessary. This is not necessarily true of all antibiotics: gentamicin CSF trough levels must be maintained below 5–20 mg/L [71], and betalactams should not be given by the intraventricular route due to neurotoxicity [72].

Though the general principle is that successful management of shunt infection can be best achieved by shunt removal, linezolid, an oxazolidinone antibiotic, might offer some prospect of retaining an infected functioning shunt due to its excellent CSF penetration and its anti-biofilm activity. An in vitro study has shown that linezolid in concentrations achievable in CSF can eradicate staphylococcal biofilms, including those of MRSA, from shunt catheters [73]. Linezolid gives high CSF levels even after oral administration [74, 75] and it has been used successfully in a small number of cases of shunt infection. Success has been achieved against vancomycin partially – resistant MRSA in shunt infection with shunt removal [76, 77]. In some cases it has been used without shunt removal. One case due to meticillin-resistant CoNS responded to oral linezolid without shunt removal after failed therapy with IV vancomycin and cefotaxime [78], and a further two, one due to MRSA and the other to meticillin-resistant CoNS, were initially unsuccessfully treated with IV vancomycin and ceftriaxone, but responded without shunt removal after IV linezolid. Further trials of this mode of management are urgently needed.

An exception to the rule that shunt removal should be the management of choice applies to those with a shunt who contract community-acquired meningitis, due to *S pneumoniae*, *H influenzae* or *N meningitidis*. As noted above, these are not true shunt infections, and the bacteria appear unable to colonise the shunt in the same way as other organisms. Clinical experience has shown that a conservative approach consisting of usual treatment for meningitis is almost always successful, and the patients usually recover quickly [79–83].

5.2 EVD infections

The general principles of treatment of shunt infections apply to EVD-associated ventriculitis, in that bacterial biofilms are involved and there may be difficulty in achieving sufficiently high CSF antibiotic levels during IV administration. An important difference from shunt infections is the much higher proportion of infections caused by gram negative bacteria, many of which are multi – drug - resistant. These include *K pneumoniae*, *A baumannii* and *Pseudomonas aeruginosa*, though the last are less common.

As soon as a diagnosis of certain or probable ventriculitis is made the EVD catheter should be removed. Failure to remove the infected EVD catheter was identified as a significant risk for treatment failure and mortality [84]. Once a new EVD catheter has been placed, appropriate antibiotic treatment should begin, and this should be guided by laboratory identification and antibiotic susceptibilities. IV colistin does not reliably result in sufficient CSF levels [85]. Again the question of IV or intraventricular administration or both arises, but the EVD makes intraventricular administration (IVT) easier. Recent guidelines from the Neurocritical Care Society [86] recommend the use of IVT "...in patients who fail to respond to IV antimicrobials alone or when organisms have high MICs to antimicrobials that do not achieve high CSF concentrations, especially MDR organisms." In a study of 31 cases, caused mainly by *Enterobacter* spp., *Ps aeruginosa* or *Stenotrophomonas* (*Xanthomonas*) *maltophilia*, the 13 cases which received IVT gentamicin as well as IV antibiotics had a higher cure rate and a lower relapse rate (0/13 vs. 6/18) [71]. None of these cases were due to *A baumannii*, which is often susceptible only to colistin. In a small study of two groups of nine patients with ventriculitis due to *A baumannii*, one group had both IV and IVT colistin while the other had IV only [87]. CSF sterilisation was achieved in 100% of those having IVT therapy (vs 33%) and the five deaths due to ventriculitis occurred in the IV - only group but none in the IV + IVT group. Cure is also reported when IVT colistin is used without IV colistin [88], so avoiding some of the systemic toxicity. Some strains of *A baumannii* are now resistant to colistin, but there is in vitro evidence of useful synergy with rifampicin, which suppressed emergence of colistin resistance as well as killing of a colistin-resistant strain. A combination of colistin and rifampicin was effective against colistin – resistant strains of *K pneumoniae* in vitro [89]. There are a few clinical reports of synergy with rifampicin, and especially if the mutual protection against resistance can be confirmed, this might improve prospects of treatment.

6. Risk factors and prevention

6.1 Shunts

Most analyses of risks for shunt infection identify shunting below the age of 1 year as a factor [8, 14, 90]. Young age or prematurity at shunting and intraventricular haemorrhage have been identified on univariate analysis but only young age on multivariate analysis, suggesting that age was the factor and the other two were dependent factors [91]. Why this should be has been debated. The main source of shunt pathogens being the patient's skin, any factor that influences this adversely might be expected to increase the risk. Premature infants have a high risk of intraventricular haemorrhage, and they often have been in hospital separated from their mothers before shunting, and it has been found that their skin bacterial densities were significantly higher, and that their skin bacteria were more likely to be able to adhere to silicone [8]. Loss of close maternal contact means that the babies become colonised with hospital strains of staphylococci that appear to be more virulent as shunt pathogens.

CSF protein content at the time of shunt insertion has been suspected to be a risk factor for infection, but this has been discounted [92, 93] though it may indicate a higher risk of re-infection after treatment of an initial shunt infection [94], possibly suggesting an incomplete eradication of the initial infection.

Intra-operative interventions to reduce the risk of shunt infection include "bundles" which are widely recognised in infection control and prevention to be effective if correctly applied. Choux et al. [95] introduced a protocol consisting of a range of sixteen measures such as restricting the number of people in the operating theatre to four, shunt insertion first thing in the morning, neonates before older children, limiting duration of surgery to 20–40 minutes, as well as technical surgical stipulations regarding a no-touch technique for the shunt, haemostasis and wound closure. On applying this to 1197 procedures he reduced the infection rate to 0.17%. Similar measures have been introduced such as use of a dedicated neurosurgical theatre, all passage into and out of the theatre prohibited, and no more than seven people present. Skin preparation used two separate applications, and both drapes and the peritoneal catheter introducers were smeared with povidone iodine which was also used to irrigate the incision and the surgeon's gloves. Using this protocol the infection rate was 0.57% [96]. The Hydrocephalus Clinical Research Network has also published protocols aimed at reducing shunt infection [97]. The refined protocol has eight essential steps, reduced from eleven in earlier versions. Results of 1935 procedures at eight centres showed an overall infection rate of 6% with 77% compliance with the protocol. Infection rates differed significantly between centres in full compliance and those which were not (5% vs. 8.7%, p = 0.005). Others have used similar protocols [98, 99]. One problem with these protocols is that they are difficult to compare, and almost none of the measures are evidence-based. However, this is considered acceptable in view of the usual fall in infection rate when they are introduced. An important aspect of bundles is mentioned by Choux [95] and emphasised by Choksey [96]: to be fully effective they must be made compulsory and violations must be detected and remedied. Though many components of the protocols are "common sense" measures such as rigorous asepsis, their main mechanism might be behavioural change in personnel, and this is not necessarily teachable and exportable to other institutions. Interestingly, few "bundles" mention the possible use of laminar flow ventilation in the OR. Choksey [96] used a laminar flow hood, while Pirotte [98] did not: both reduced their shunt infection rate to <1%. Though laminar flow ventilation was recommended for arthroplasty, recently several centres have reported either no benefit, or in some cases a small but significant increase in infection rate [100, 101].

However, certain constituent measures deserve attention. Many use antiseptics to either irrigate the incision or to isolate the wound skin edges [102], a measure suggested some time ago [103–105]. It is clear that skin bacteria enter the incision from this source [27, 106]. It is important that the contribution of patient skin bacteria is minimised by avoiding contact with skin edges by gloves, instruments or shunt components, and measures to isolate them might be helpful in this regard. Surgeons' gloves become contaminated early in the operation and double – gloving is recommended so that the contaminated outer pair can be discarded before the shunt is handled. At this point it is advisable to rinse the gloved hands in antiseptic before touching the shunt, or to use a "no - touch" technique. Double gloving was introduced in a sequential study but without removal of the outer pair, as the presumption was that the source of shunt pathogens was glove perforation [107]. However, the diagnostic criteria in this study are in doubt as most of the "infections" were culture-negative, and the change of outer glove remains the most important measure.

It is important to remember that, irrespective of the antiseptic used, pre-operative skin preparation does not sterilise the skin. Much of the literature discusses the merits of various antiseptics but relies on skin swabs for evaluation, though most of the skin flora reside in the glands and follicles in the dermis [108]. When full thickness skin biopsies have been used they have shown that, irrespective of the agent used, while the numbers of bacteria can be reduced they cannot be eradicated, and they will re-emerge during surgery. The numbers of bacteria required to cause an infection in the presence of a biomaterial such as a shunt are at least ten thousand – fold fewer that those needed in its absence [109]. While there is little evidence that chlorhexidine is better than povidone iodine it is clear that the alcohol version of either is superior to the aqueous version, and it is possible that the alcohol component is the major antiseptic factor [110].

Adhesive drapes are not of proven benefit in preventing the skin bacteria from entering the incision, even when iodine-treated. They are, however, useful in covering the cloth drapes and providing a dry aseptic surface. Shaving of head hair is now accepted as unnecessary and possible a risk for infection [111], and clipping should be carried out with scissors if necessary, and the hair prepared as for the skin.

The use of pre-operative prophylactic antibiotics is controversial [112]. Most of the reports, including where infection rates are considered unacceptably high, are from centres using antibiotic prophylaxis. Again the issue of timely penetration of IV antibiotics into the CSF is important, and most studies have found ineffective peri-operative levels. They also do not act rapidly enough to affect the numbers of skin bacteria in the incision during shunt insertion, and while this remains as a risk to the shunt, they might act to reduce postoperative wound infection. Intra-operative IVT vancomycin has not been shown to reduce shunt infection rate, probably due to the slow kill rate, but an interesting finding has been reported [113]. In this study, when only IV antibiotics were used, the infection rate was 6.74%; when IVT gentamicin was added, the infection rate was similar at 5.45%; when both IVT gentamicin and IVT vancomycin were used together, the infection rate fell to 0.41%. This interesting observation needs to be confirmed.

Another approach is the use of triclosan - coated sutures [114]. Though numbers of patients were small, when triclosan - coated sutures were compared with plain sutures in a randomised controlled trial, there was a reduction in shunt infection rate from 21% to 4.3%. Some, but not all, infections were postoperative wound suppurations.

There is increasing use of topical application of vancomycin powder before fascial closure in spinal surgery with significant infection reduction and low toxicity compared to IV vancomycin prophylaxis. The same approach has been used in a small uncontrolled series of shunt insertions with a reduction of shunt infection from 5.8% to 0% though the postoperative revision rate was unaltered [115]. The diagnostic criteria for shunt infection were not clear but this use of vancomycin is safe and effective in spine surgery and might be useful in shunt surgery.

It appears that attempts to prevent skin bacteria from accessing the shunt during surgery are only partly successful, and further measures are needed. Systemic prophylactic antibiotics are well researched but an unacceptably high infection rate remains. This has led to development of shunt materials intended to reduce bacterial colonisation of the catheters and therefore shunt infection.

Antimicrobial shunt catheters have been available for some time. Coating the shunt surface with a hydrophilic material as in the Bioglide catheter is intended to reduce bacterial attachment, and if the catheter is soaked in a solution of antibiotic then this has been claimed to add to the effect. The catheters have been evaluated in vitro using rigorous clinically predictive tests and though they did reduce bacterial attachment they were found not to be effective in preventing colonisation by staphylococci [116] even when soaked in gentamicin or vancomycin [117]. Clinical assessment [118, 119] has confirmed this. Silver in various forms has been promoted as a useful antimicrobial for implantable devices, but variable results have been reported. Shunt catheters impregnated with nanoparticulate silver, a particularly active form, have been marketed. Again an in vitro evaluation has found that they failed to prevent colonisation by staphylococci, *C acnes* or *E coli*, and this has been confirmed in a large randomised controlled clinical trial [6]. Silver undoubtedly has antibacterial activity, but the concentrations of silver ions needed are also cytotoxic, and silver ions combine avidly with chloride and protein. As it is likely that a prolonged duration of antimicrobial activity of at least a few days is required to prevent survival and regrowth of bacteria in shunt catheters, and as antimicrobial coatings are easily removed by CSF flow and obliterated by protein deposition, a system is needed that maintains an antimicrobial surface. One such system distributes molecules of antimicrobials throughout the silicone matrix, allowing them to migrate freely to replenish the surface when CSF removes molecules from there, so maintaining an antimicrobial surface for sufficient time, in this case for over 40 days [120]. This system is unaffected by protein. When the clinically predictive tests are applied in vitro, the antimicrobial catheters remain free of bacterial colonisation even after serial high - dose bacterial challenge. Clinical studies have demonstrated reduction in shunt infection [121] and considerable cost savings [122] as well as reduction in systemic antibiotic use. A large randomised controlled trial comparing antimicrobial, silver and plain shunts found that the antimicrobial shunts gave a statistically significant reduction in shunt infections while results for silver-processed shunts were indistinguishable from those of plain catheters [6].

The question of whether to use prophylactic antibiotics for people with shunts who undergo dental treatment has been raised frequently. Studies have shown that the risk is negligible, and there is no evidence that bacteria of oral origin have caused shunt infections, whether VP or VA, after dental treatment [123]. It is likely that antibiotic prophylaxis used in this way treats the dental practitioner but is of no benefit to the patient.

6.2 EVDs

Whether risks of infection are different if the EVD is inserted in the intensive care unit or the operating room (OR) is debatable. In one study there were fewer infections in those inserted in the OR but this difference was not statistically significant [124]. While the OR might offer a more controllable aseptic environment, transfer of acutely ill patients to the OR for this purpose might pose additional risks [125].

Periprocedural prophylactic antibiotics are commonly used. However, many centres use prolonged antibiotic prophylaxis throughout the EVD use. Flibotte et al. [126] used either nafcillin or vancomycin but noted that most infections were still due to gram positive pathogens, and that their use of prolonged nafcillin appeared to lead to an increase in resistance to this antibiotic. Wong and Poon [127] reported a comparison of two regimens for prolonged prophylaxis but did not comment on the influence on resistance, but a later study by the same authors [128] using the same regimen found three cases of pseudomembranous colitis due to *Clostridioides difficile*, one of whom required total colectomy. A similar experience was reported [129] reducing the number of *C difficile* infections in the ICU from 19 to 5 by changing the antibiotic prophylaxis regimen from prolonged to peri-procedural with no change in ventriculitis rate. Antimicrobial impregnated EVD catheters were used in both phases of both these studies [128, 129]. In another study comparing prolonged and peri-procedural antibiotics there was no difference in infection rate, the difference being a saving of $80,000 a year in drug costs [130]. Murphy et al. [131] also compared a period when prolonged antibiotics were used with a period where only periprocedural antibiotics were given; in both periods antimicrobial EVD catheters were used. The infection rate in the periprocedural-only period actually fell from 1.35/1000 catheter days to 0.54/1000 catheter days, though this was not statistically significant. Remarkably, there was a significantly higher rate of bloodstream infections (BSI) and pneumonia (VAP) in the prolonged – antibiotics period, and the drug cost for treating these infections were $155,253 but there were no cases of BSI or VAP in the second periprocedural - only period.

Antimicrobial catheters have been developed for EVDs as for shunts, though there have been fewer clinical trials. The hydrophilic-coated catheters intended to reduce infection by preventing bacterial attachment have already been discussed; these have not been successful. Silver-processed catheters have shown non-significant results in some clinical studies [132–135]. One three-phase retrospective/ sequential study showed that introduction of silver-processed catheters reduced the infection rate from 3.8% to zero, though due to small numbers this was not statistically significant [136]. A randomised prospective controlled trial comparing silver-processed with plain catheters has reported a significant reduction in ventriculitis from a very high rate of 21.4% to 12.3%, a fall that just met statistical significance $p = 0.0427$ [137]. A thorough assessment of the value of silver-processed EVD catheters [138] has found no significant overall difference in infection rate in a meta-analysis but did identify a statistically significant reduction in infection due to gram positive bacteria (6.7–2%, p = 0.002). The conclusion was that silver-processed catheters require further evaluation, and that they have no activity against gram negative bacteria. This was confirmed in vitro using the same rigorous clinically – predictive testing used for shunts, when silver-processed EVD catheters were found to show a weak activity against *S epidermidis* but none against gram negative bacteria [139].

An antibiotic-impregnated catheter containing rifampicin and minocycline (VentriClear, Cook Inc) is available in USA, and Bactiseal (Codman Integra Life Sciences) that contains rifampicin and clindamycin, produced by a different process, is available worldwide. A significant (p = 0.002) reduction in ventriculitis has been reported when VentriClear catheters were used, from 9.4% to 1.3% [140]. Harrop et al. [12] carried out a five -phase prospective cohort study. Phase I, the baseline, showed a rate of 6.7%, and the introduction of a standardised protocol in Phase II did not reduce this (8.2%). However, in Phase III the Bactiseal catheter was included, and the infection rate fell to 1% (p = 0.0005). This catheter gave an unacceptable rate of occlusion and its use was discontinued, and reversion to the Phase II protocol showed a return to a 7.6% rate. In Phase V, the VentriClear catheter was introduced and the ventriculitis rate again fell to 0.9% (p = 0.0001). Though a sequential cohort study, this provided strong evidence that both antimicrobial catheters were effective. This was confirmed by a comparison of VentriClear with Bactiseal [141] using alternating 3-month periods when 129 patients received either a VentriClear or a Bactiseal catheter. No cases of ventriculitis were recorded in the study, showing that both were equally effective in this study. No excess of occlusion was recorded with either catheter. A series involving historical controls found a reduction of ventriculitis from 15% in plain catheters to 5% in Bactiseal catheters but this failed to reach statistical significance [142]. Bactiseal was compared with plain EVD catheters in an interesting study in which CSF samples were taken every 2 days, and if culture-positive, irrespective of clinical evidence, the catheter was changed and 10 days of intraventricular antibiotics were given [143]. In this study there were no cases of clinical infection in either group. As with shunts, there is no evidence that antimicrobial-impregnated EVD catheters increase the risk of bacterial resistance, and in reducing the need for systemic antibiotics for prophylaxis and treatment of infections they are likely to contribute to reduction of antimicrobial resistance generally. This has been underlined by three studies in which prolonged systemic antibiotic prophylaxis for EVD has been compared with use of the Bactiseal catheter. In two studies the antimicrobial catheter gave equivalent protection against ventriculitis but avoided the serious risk of *C difficile* infection [128, 129], a known consequence of over-use of antibiotics. In the third study [131] the Bactiseal EVD catheter was used but in one group, prolonged antibiotic prophylaxis were added. There was no difference in ventriculitis rate between the two groups, but there was a significantly higher rate of BSI and VAP, requiring further courses of antibiotics for treatment, again contributing to antimicrobial resistance. While good quality randomised controlled trials are needed for antimicrobial - impregnated EVD catheters, the studies so far strongly suggest that they can reduce the incidence of ventriculitis by gram positive bacteria, but there is currently no EVD catheter available that protects against gram negative bacteria, which are increasing in frequency and importance. An experimental antimicrobial EVD catheter that can protect against colonisation by MDR gram negative bacteria including *A baumannii* has been developed but is not yet clinically available [144].

There is general agreement that the EVD catheter must be tunnelled subcutaneously for approximately 5 cm away from the burr hole, but some prefer to tunnel for much longer. When the exit site was placed in the lower chest or upper abdomen, no infections were reported in the first 16 days. In those 45 requiring EVD for longer, four developed ventriculitis [145]. In a study using a long tunnel of at least 20 cm an infection rate significantly lower than those reported in the literature using conventional tunnels was noted, though an antimicrobial EVD catheter was also used [146]. However, Leung et al. [147] found no advantage in a longer tunnel.

The infection rate for EVDs is said to rise with duration of use, though this is sometimes contested. The duration of EVD use has frequently been identified as a risk factor for infection. As the increase in infection appears to begin after about 5 days, suggestions have been made that changing the EVD catheter at this stage might avoid the subsequent rise in infection rate [148]. However, this practice has been shown not to help [149, 150] and might increase the risk [151]. The risk for ventriculitis increases in most studies until about 10–12 days then levels off. The message is that the EVD should be removed as soon as possible when no longer needed.

EVD pathogens are more varied, and more likely to be MDR gram negative bacteria than those found in shunts. They might originate on the patient's body surfaces, ears and respiratory tract as broad-spectrum antibiotics given for chest infections and other purposes promote colonisation of these sites with such organisms. The intensive care environment is often a source of such bacteria due to the throughput of very sick patients and heavy use of antibiotics. This environment includes all inanimate surfaces, textiles and water sources [152, 153]. The EVD must be managed with careful attention to aseptic technique, and breaches of the system should be avoided unless absolutely necessary. This includes CSF sampling for monitoring purposes. The practice of daily CSF sampling is said to enable early diagnosis of infection [154, 155], but represents a risk for introduction of infection, and CSF sampling is best confined to cases where there is a suspicion of infection [155].

As with shunts, the introduction of "bundles" has usually been associated with a reduction in infection rate. Korinek et al. [156] developed a bundle protocol and introduced a violation score to monitor it. Their ventriculitis rate fell from 9.9% to 4.6%, and the most significant factors in infected patients were CSF leak and protocol violation, which in the infected cases was 4 times higher $p < 0.0001$. The value of this approach was also demonstrated by others with a significant fall in infection rate [155, 157]. Importantly, the bundle approach should include full involvement of all personnel involved and should be monitored and regular feedback given on violations and infection rates. Again, not all of the constituents of the bundle are evidence – based and they vary between reports, but the behaviour change brought about by this approach is the most important factor.

7. Conclusions

Infection in shunting and EVD is often devastating. Prevention is paramount and a greater understanding of the science and the risk factors should inform more effective measures including surgical practice and OR discipline. Antimicrobial catheters are useful in reducing infection in shunts and EVDs, but the problem of gram negative infection needs to be addressed. There should be no delay in instituting effective treatment, including removal of hardware and ensuring adequate levels of antibiotics. Successful first pass treatment should be the goal. Treatment without hardware removal, using relatively new antibiotics, should be thoroughly investigated in view of the potential benefits.

Importantly, the contribution of overuse or misuse of antibiotics to the increasing problem of antimicrobial resistance both locally and globally should be kept in mind.

Conflict of interest

The author is the inventor of the "Bactiseal" antimicrobial catheter, but he has not and does not receive any royalties or other payment. He receives speaker fees from Codman Inc., but not for personal gain and these are paid to his University.

Author details

Roger Bayston
Academic Unit of Injury, Inflammation and Recovery Science, School of Medicine, University of Nottingham, Nottingham, United Kingdom

*Address all correspondence to: roger.bayston@nottingham.ac.uk

References

[1] Carrington KW. Ventriculovenous shunt using the Holter valve as a treatment of hydrocephalus. J Michigan Med Soc. 1959;58:373-376.

[2] Holt RJ. The classification of staphylococci from colonised ventriculo-atrial shunts. J Clin Pathol.1969;22:475-482.

[3] Ingraham FD, Matson DD, Alexander E, Woods RP. Studies in the treatment of experimental hydrocephalus. J Neuropath Exp Neurol. 1948;7:123-143

[4] Shurtleff DB, Folz EL, Christie D. Ventriculoauriculostomy-associated infection: a 12-year study. J Neurosurg. 1971;35:686-694.

[5] Schoenbaum SC, Gardner P, Shillito J. Infections of cerebrospinal fluid shunts: epidemiology, clinical manifestations and therapy. J Infect Dis. 1975;131:543-552.

[6] Mallucci CL, Jenkinson MD, Conroy EJ, Hartley JC, Brown M, Dalton J, Kearns T, Moitt T, Griffiths MJ, Culeddu G, Solomon T, Hughes D, Gamble C. Antibiotic or silver versus standard ventriculoperitoneal shunts (BASICS): a multicentre, single-blinded, randomised trial and economic evaluation. Lancet. 2019;394:1530-1539. doi.org/10.1016/S0140-6736(19) 31603-4

[7] Renier D, Lacombe J, Pierre-Kahn A, Sainte-Rose C, Hirsch J-F. Factors causing acute shunt infection. J Neurosurg. 1984;61:1072-1078.

[8] Pople IK, Bayston R, Hayward RD. Infection of cerebrospinal fluid shunts in infants: a study of etiological factors. J Neurosurg. 1992;77:29-36.

[9] Kulkani AV, Drake JM, Lambert-Pasculli M. Cerebrospinal fluid shunt infection: a prospective study of risk factors. J Neurosurg. 2001;94: 195-201.

[10] Schade RPS, Schinkel J, Visser LG, van Dijk JMC, Voormolen JHCV, Kuijper EJK. Bacterial meningitis caused by the use of ventricular or lumbar cerebrospinal fluid catheters. J Neurosurg. 2005;102:229-234.

[11] Hoefnagel D, Dammers R, ter Laak-Poort MP, Avezaat CJJ. Risk factors for infections related to external ventricular drainage. Acta Neurosurg (Wien). 2008;150:209-214. DOI 10.1007/s00701-007-1458-9

[12] Scheithauer S, Bürgel U, Bickenbach J, Häfner H, Haase G, Waitschies B, Reinges MHT, Lemmen SW. External ventricular and lumbar drainage -associated meningoventriculitis: prospective analysis of time-dependent infection rates and risk factors. Infection. 2010;38:205-209. DOI: 10.1007/s15101-010-0006-3

[13] Harrop JS, Sharan AD, Ratcliff J, Prasad S, Jabbour P Evans JJ, Veznedaroglu E, Andrews DW, Maltenfort M, Liebman K, Flomenberg P, Sell B, Baranoski AS, Fonshell C, Reiter D, Rosenwasser RH. Impact of a standardised protocol and antibiotic – impregnated catheters on ventriculostomy infection rates in cerebrovascular patients. Neurosurg. 2010;67:187-191.

[14] Lee JK, Seok JY, Lee JH, Choi EH, Phi JH, Kim SK, Wang KC, Lee HJ. Incidence and risk factors of ventriculoperitoneal shunt infections in children: a study of 333 consecutive shunts in 6 years. J Korean Med Sci. 2012;27:1563-1568

[15] James G, Hartley JC, Morgan RD, Ternier J. Effect of introduction of

antibiotic – impregnated shunt catheters on cerebrospinal fluid shunt infection in children: a large single-center retrospective study. J Neurosurg Pediatr. 2014;13:101-106

[16] Farber SH, Parker SL, Adogwa O, McGirt MJ, Rigamonti D. Effect of antibiotic-impregnated shunts on infection rate in adult hydrocephalus: a single institution's experience. Neurosurg. 2011;69:625-629.DOI: 10.1227/NEU.0b013e31821bc435

[17] Lee MJ, Pottinger PS, Butler-Wu S, Bumgarner RE, Russ SM, Matsen FA. Propionibacterium persists in the skin despite standard surgical preparation. J Bone Joint Surg. 2014;96:1447-1450. DOI: org/10.2106/jbjs.m.01474

[18] Arnell K, Cesarini K, Lagerqvist-Widh A, Wester T, Sjölin J. Cerebrospinal fluid shunt infections in children over a 13-year period: an-aerobic cultures and comparison of clinical signs of infection with *Propionibacterium acnes* and with other bacteria. J. Neurosurg. Pediatr. 2008;**1:**366-372.

[19] Thompson TP, Albright AL. *Propionibacterium acnes* infections of cerebrospinal fluid shunts. Child's Nerv Syst. 1998;14:378-380

[20] Conen A, Walti LN, Merlo A, Fluckiger U, Battegay M, Trampuz A. Characteristics and treatment outcome of cerebrospinal fluid shunt-associated infections in adults: a retrospective analysis over an 11 -year period. Clin Infect Dis. 2008;47:73-82. DOI: 10.1086/588298

[21] Arnell K, Enblad P, Wester T, Sjölin J. Treatment of cerebrospinal fluid infections in children using systemic and intraventricular antibiotic therapy in combination with externalization of the ventricular catheter: efficacy in 34 consecutively treated infections. J Neurosurg Pediatr.2007;107:213-219

[22] Park J, Choi Y-J, Ohk B, Chang H-H. Cerebrospinal fluid leak at percutaneous exit of ventricular catheter as a crucial risk factor for external ventricular drainage-related infection in adult neurosurgical patients. World Neurosurg. 2018;109:e398-e403. DOI. org/10.1016/j.wneu.2017.09.190

[23] Walti LN, Conen A, Coward J, Jost GF, Trampuz A. Characteristics of infections associated with external ventricular drains of cerebrospinal fluid. J Infect. 2013;66:424-431. DOI: 10.1016/j.jinf.2012.12.010

[24] Chatzi M, Karvouniaris M, Makris D, Tsimitrea E, Gatos C, Tasou A, Manzarlis K, Zakinthinos E. Bundle of measures for external cerebral ventricular drainage-associated ventriculitis. Crit Care Med. 2014;41:66-73. DOI: 10.1097/CCM.0b013e3182 9a70a5

[25] Lyke KE, Obasanjo OO, Williams MA, O'Brien M, Chotani R, Peri TM. Ventriculitis complicating use of intraventricular catheters in adult neurosurgical patients. Clin Infect Dis. 2001;33:2028-2033.

[26] Chi H, Chang K-Y, Chang H-C, Chiu, N-C, Huang F-Y. Infections associated with indwelling ventriculostomy catheters in a teaching hospital. Internat J Infect Dis. 2010;14:e216-e219. DOI: 10.1016/j. ijid.2009.04.0006

[27] Bayston R, Lari J.A study of the sources of infection in colonised shunts. Dev Med Child Neurol. 1974;32:16-22

[28] LeVine SM, Wulser MJ, Lynch SG. Iron quantification in cerebrospinal fluid. Anal Biochem. 1998;265:74-78. DOI: 10.1006/abio.1998.2903

[29] Bayston R, Penny SR. Excessive production of mucoid substance in Staphylococcus SIIA: a possible factor in colonisation of Holter shunts. Dev Med Child Neurol. 1972;14:25-28

[30] Guevara JA, Zuccaro G, Trevisan A, Denoya CD. Bacterial adhesion to cerebrospinal fluid shunts. J Neurosurg. 1987;67:438-445.

[31] Fux CA, Quigley M, Worel AM, Post C, Zimmerli S, Ehrlich G, Veeh RH. Biofilm-related infections of cerebrospinal fluid shunts. Clin Microbiol Infect. 2006;12:331-337.

[32] Darouiche RO, Dhir A, Miller AJ, Landon GC, Raad II, Musher DM. Vancomycin penetration into biofilm covering infected prostheses and effect on bacteria. J Infect Dis.1994;170: 720-723.

[33] Gilbert P, Maira-Litran T, McBain AJ, Rickard AH, Whyte FW. The physiology and collective recalcitrance of microbial biofilm communities. Adv. Microb. Physiol. 2002;46:202-256.

[34] Duguid IG, Evans E, Brown MRW, Gilbert P. Effect of biofilm culture upon the susceptibility of *Staphylococcus epidermidis* to tobramycin. J Antimicrob Chemother. 1992;30, 803-810.

[35] Anwar H, Costerton JW. Effective use of antibiotics in the treatment of biofilm-associated infections. *ASM News Journal* 1992;58, 665-668.

[36] Choi AHK, Slamti L, Avci FY, Pier G, Maira-Litran T. The *pgaABCD* Locus of *Acinetobacter baumannii* Encodes the Production of Poly-beta-1-6-*N*-Acetylglucosamine, Which Is Critical for Biofilm Formation. J Bacteriol. 2009;191:5953-5963. DOI: 10.1128/JB.00647-09

[37] Bayston R, Ashraf W, Barker-Davies R, Tucker E, Clement R, Clayton J, Freeeman BJC, Nuradeen B. Biofilm formation by *Propionibacterium acnes* on biomaterials in vitro and in vivo: impact on diagnosis and treatment. J Biomed Mater Res A. 2006;81:705-709. DOI: 10.1002/jbm.a.31145

[38] Ghritlaharey RK, Budhwani KS, Shrivastava DK, Gupta G, Kushwaha AS, Chanchlani R, Nanda M. Trans-anal protrusion of ventriculo-peritoneal shunt catheter with silent bowel perforation: report of ten cases in children. Pediatr Surg Int. 2007;23:575-580.

[39] Bayston R, Spitz L. Infective and cystic causes of malfunction of ventriculoperitoneal shunts. Zeit Kinderchirurg. 1977;22,419-424.

[40] Tamura A, Shida D, Tsutsumi K. Abdominal cerebrospinal fluid pseudocyst occurring 21 years after ventriculoperitoneal shunt placement: a case report. BMC Surg. 2013;13:27. DOI: 10.1186/1471-2482-13-27

[41] Bayston R, Swinden J.The aetiology and prevention of shunt nephritis. Z Kinderchir. 1979.28:377-384

[42] Haffner D, Schinderer F, Aschoff A, Matthias S, Waldherr R, Schärer K. The clinical spectrum of shunt nephritis. Nephrol Dial Transplant. 1997;12:1143-1148. DOI: 10.1093/ndt/12.6.1143

[43] ter Borg EJ, van Rijswijk MH, Kallenberg CG. Transient arthritis with positive tests for rheumatoid factor as presenting sign of shunt nephritis. Ann Rheum Dis. 1991;50:182-183.

[44] Legoupil N, Ronco P, Berenbaum F. Arthritis-related shunt nephritis in an adult. Rheumatol. 2003;42:698-699.

[45] Vella J, Carmody M, Campbell E, Browne O, Doyle G, Donohoe J. Glomerulonephritis after ventriculo-atrial shunt. Q Med J. 1995;88:911-918.

[46] Ben-Ami R, Navon-Venezia S, Schwartz D, Carmeli Y. Infection of a Ventriculoatrial Shunt with Phenotypically Variable *Staphylococcus epidermidis* Masquerading as Polymicrobial Bacteremia Due to Various Coagulase-Negative Staphylococci and *Kocuria varians.* J Clin Microbiol. 2003; 2444-2447. DOI: 10/1128/jcm.41,6,2444-2447.2003

[47] Spanu T, Romano L, D'Inzeo T, Masucci L, Albanese A, Papacci F, Marchese E, Sanguinetti M, Fadda G. Recurrent ventriculoperitoneal shunt infection caused by small-colony variants of *Staphylococcus aureus*. Clin Infect Dis. 2005;41:48-52.

[48] Holt RJ. The early serological detection of colonisation by *Staphylococcus epidermidis* of ventriculoatrial shunts. Infection. 1980;8:8-12.

[49] Clayton J, Bayston R, Donald F. Occult ventriculo-atrial shunt infection: a forgotten condition. Cerebrospinal Fluid Res. 2005;2(suppl 1):S23. doi:10.1186/1743-8454-2-S1-S23

[50] Muttaiyah S, Ritchie S, Upton A, Roberts S. Clinical parameters do not predict infection in patients with external ventricular drains: a retrospective observational study of daily cerebrospinal fluid analysis. J Med Microbiol. 2008;57:207-209. DOI 10.1099/jmm.0.47518-0

[51] Beer R, Lackner P, Pfausler B, Schmutzhard E. Nosocomial ventriculitis and meningitis in neurocritical care patients. J Neurol. 2008;255:1617-1624.

[52] Pfausler B, Beer R, Engelhardt K, Kemmler G, Mohsenipour I, Schmutzhard E. Cell index- a new parameter for the early diagnosis of ventriculostomy (external ventricular drainage) – related ventriculitis in patients with intraventricular hemorrhage? Acta Neurochir (Wien). 2004;146:477-481. DOI 10.1007/s00701-004-0258-8

[53] Gordon M, Ramirez P, Soriano A, Palomo M, Lopex-Ferraz C, Villareal E,Meseguer S, Gomez MD, Folgado C, Bonatre J. Diagnosing external ventricular drain-related ventriculitis by means of local inflammatory response: soluble triggering receptor expressed on myeloid cells-1. Crit Care. 2014;18:567 DOI.org/10.1186/s13054-014-0567-0

[54] Pfisterer W, Mühlbauer M, Czech T, Reinprecht A. Early diagnosis of external ventricular drainage infection. Results of a prospective study. J Neurol Neurosurg Psychiatr. 2003;74:929-932. DOI: 10.1136/jnnp.74.7.929

[55] Schade RP, Schinkel J, Roelandse FW, Geskus RB, Visser LG, van Dijk JM, Voormolen JHC, van Pelt H, Kuijper EJ. Lack of value of routine analysis of cerebrospinal fluid for prediction and diagnosis of external drainage-related bacterial meningitis. J Neurosurg. 2006;104:101– 108. DOI: 10.3171/jns.2006.104.1.101

[56] Tunkel AR, Hasbun R, Bhimraj A, Byers K, Kaplan SL, Scheld WM, van der Beek D, Bleck TP, Garton HJL, Zunt JR. 2017 Infectious Diseases Society of America's Clinical Practice Guidelines for healthcare-associated ventriculitis and meningitis. Clin Infect Dis. 2017;64:701-706. doi.org/10.1093/cid/ciw861

[57] Lutsar I, McCracken GH, Friedland IR. Antibiotic pharmaco dynamics in cerebrospinal fluid. Clin Infect Dis. 1998;27:1117-1129.

[58] James HE, Walsh JW, Wilson HD, Connor JD, Bean JR, Tibbs PA. Prospective randomized study of therapy in cerebrospinal fluid shunt infection. Neurosurg. 1980;7:459-463.

[59] James HE, Bradley JS: Aggressive management of shunt in- fection: combined intravenous and intraventricular antibiotic therapy for twelve or less days. Pediatr Neurosurg. 2008;44:104– 111.

[60] Schreffler RT, Schreffler AJ, Wittler RR. Treatment of cerebrospinal fluid shunt infections: a decision analysis. Pediatr Infect Dis J. 2002;21:632-636.

[61] Sullins AK, Abdel-Rahman SM. Pharmacokinetics of antibacterial agents in the CSF of children and adolescents. Pediatr Drugs. 2013;15:93-117. DOI: 10.1007/s40272-013-0017-5

[62] Edwards MS, Baker CJ, Butler KM, Mason EO, Laurent JP, Cheek WR. Penetration of cefuroxime into ventricular fluid in cerebrospinal fluid shunt infections. Antimicrob Ag Chemother. 1989;33:1108-1110.

[63] Jorgensen L, ReiterPD, Freeman JE, Winston KR, Fish D, McBride LA, Handler MH. Vancomycin Disposition and Penetration into Ventricular Fluid of the Central Nervous System following Intravenous Therapy in Patients with Cerebrospinal Devices. Pediatr Neurosurg. 2007;43:449-455. DOI: 10.1159/000108786

[64] Pfausler B, Spiss H, Beer R, Kampfl A, Engelhardt K, Schober M, Schmutzhard E. Treatment of staphylococcal ventriculitis associated with external cerebrospinal fluid drains: a prospective randomized trial of intravenous compared with intraventricular vancomycin therapy. J Neurosurg. 2003;98:1040-1044.

[65] Nau R, Blei C, Eiffert H. Intrathecal antibacterial and antifungal therapies. Clin Microbiol Rev. 2020;33e00190-19. DIO.org/10.1128/CMR.00190-19

[66] Brown EM, de Louvois J, Bayston R, Hedges AJ, Johnston RA, Lees P. Infection in Neurosurgery Working Party of the British Society for Antimicrobial Chemotherapy. Antimicrobial prophylaxis in neurosurgery and after head injury. Lancet 1994;344:1547 -1551.

[67] Bayston R, Hart CA, Barnicoat M. Intraventricular vancomycin in the treatment of ventriculitis associated with cerebrospinal fluid shunting and drainage. J Neurol Neurosurg Psychiatr. 1987;50:1419-1423.

[68] Thompson JB, Einhaus S, Buckingham S, Phelps SJ. Vancomycin for treating cerebrospinal fluid shunt infections in pediatric patients. J Pediatr Pharmacol Ther. 2005;10:14-25.

[69] Beach JE, Perrott J, Turgeon RD, Ensom MHH. Penetration of vancomycin into the cerebrospinal fluid: a systematic review. Clin Pharmacokinet. 2017;56:1479-1490. DOI: 10.1007/s40262-017-1548-y

[70] Kossman T, Hans V, Stocker R, Imhof H-G, Joos B, Trentz O, Morganti-Kossman MC. Penetration of cefuroxime into the cerebrospinal fluid of patients with traumatic brain injury. J Antimicrob Chemother. 1996;37: 161-167.

[71] Tängden T, Enblad P, Ullberg M, Sjölin J. Neurosurgical Gram-negative bacillary ventriculitis and meningitis: a retrospective study evaluating the efficacy of intraventricular gentamicin therapy in 31 consecutive cases. Clin Infect Dis. 2011;52:1310-1316.

[72] Wen DY, Bottini AG, Hall WA, Haines SJ. Infections in neurologic surgery. The intraventricular use of antibiotics. Neurosurg Clin N Am. 1992;3:343-54.

[73] Bayston R, Ullas G, Ashraf W. Action of linezolid or vancomycin on biofilms in ventriculoperitoneal shunts in vitro. Antimicrob Agents Chemother.

2012;56:2842-2845. DOI: 10.1128/AAC.06326-11

[74] Diekma DI, Jones RN. Oxazolidinones: a review. Drugs. 2000;59:7-16.

[75] Gill CJ, Murphy MA, Hamer DH. Treatment of *Staphylococcus epidermidis* ventriculo- peritoneal shunt infection with linezolid. J Infect. 2002;45:129-132

[76] Amod F, Moodley I, Peer AKC, Sunderland J, Lovering A, Wooton M, Navdi S, Vawda F. Ventriculitis due to a hetero strain of vancomycin intermediate *Staphylococcus aureus* (hVISA): successful treatment with linezolid in combination with intraventricular vancomycin. J Infect. 2005;50:252-257.

[77] Cook AM, Ramsey CN, Martin CA, Pittman T. Linezolid for the treatment of a heteroresistant *Staphylococcus aureus* shunt infection. Ped Neurosurg. 2005;41:102-104. DOI: 10.1159/000085165

[78] Castro P, Soriano A, Escrich C, Villalba G, Sarasa M, Mensa J. Linezolid treatment of ventriculoperitoneal shunt infection without implant removal. Eur J Clin Microbiol Infect Dis. 2005;24:603-606. DOI 10.1007/s10096-005-0015-9

[79] Patriarca PA, Lauer BA. Ventriculoperitoneal shunt-associated infection due to *Haemophilus influenzae*. Pediatr. 1980;65:1007-1009.

[80] Rennals MB, Wald ER. Treatment of *Haemophilus influenzae* type b meningitis in children with cerebrospinal fluid shunts. J Pediatr. 1980;97:424-426.

[81] Petrak RM, Pottage JC, Harris AA, Levin S. *Haemophilus influenzae* meningitis in the presence of a cerebrospinal fluid shunt. Neurosurg. 1986; 18:79-81.

DOI:10.1227/00006123-198601000-00013

[82] Stern S, Bayston R, Hayward RJ. *Haemophilus influenza*e meningitis in the presence of cerebrospinal fluid shunts. Childs Nerv Syst. 1988;4:164-165.

[83] O'Keeffe PT, Bayston R. Pneumococcal meningitis in a child with a ventriculoperitoneal shunt. J Infect. 1991;22:77-79.

[84] Rodríguez-Lucas C, Fernández J, Martínez-Sela M, Álvarez-Vega M, Moran N, Garcia A, Menendez C, Garcia Prieto E, Rodriguez-Guardado A. *Pseudomonas aeruginosa* nosocomial meningitis in neurosurgical patients with intraventricular catheters: therapeutic approach and review of the literature. Enferm Infecc Microbiol Clin. 2020;38:54-58. DOI: 10.1016/j.eimc.2019.04.033

[85] Markantonis SL, Markou N, Fousteri M, Sakellaridis N, Karatzas S, Alamanos I, Dimipoulou E, Baltopoulos G. Penetration of colistin into cerebrospinal fluid. Antimicrob Ag Chemother. 2009;53:4907-4910. DOI: 10.1128/AAC.00345-09

[86] Fried HI, Barnett RN, Rowe AS, Zabramski JM, Andaluz N, Bhimraj A, Guanci MM, Seder DB, Singh JM. The insertion and management of external ventricular drains: an evidence-based consensus statement. Neurocrit Care. 2016;24:61-81. DOI 10.1007/s12028-015-0224-8

[87] De Bonis P, Lofrese G, Scoppettuolo G, Spanu T, Cultrera R, Labonia M, Cavallo MA, Mangiola A, Anile C, Pompucci A. Intraventricular versi]us intravenous colistin for the treatment of extensively drug – resistant *Acinetobacter baumannii* meningitis. Europ J Neurol. 2016;23:68-75. DOI: 10.1111/ene.12789

[88] Bargiacchi O, Rossati A, Car P, Brustia D, Brondolo R, Rosa F, Garavelli PL, de Rosa FG. Intrathecal/intraventricular colistin in external ventricular device-related infections by multi-drug resistant Gram negative bacteria: case reports and review. Infection. 2014;42:801-809. DOI: 10.1007/s15010-014-0618-0

[89] Tascini C, Tagliaferri E, Giani T, Leonildi A, Flammini S, Casini B, Lewis R, Ferranti S, Rossolini GM, Menichetti F. Synergistic activity of colistin plus rifampin against colistin-resistant KPC-producing *Klebsiella pneumoniae*. Antimicrob Agents Chemother. 2013;57:3990-3993.

[90] Renier D, Lacombe J, Pierre-Kahn A, Sainte-Rose C, Hirsch J-F. Factors causing acute shunt infection. J Neurosurg. 1984;61: 1072-1078.

[91] McGirt MJ, Zaas A, Fuchs HE, George TM, Kaye K, Sexton DJ. Risk factors for pediatric ventriculoperitoneal shunt infection and predictors of infectious pathogens. Clin Infect Dis. 2003;36:858-862.

[92] Brydon HL, Bayston R, Hayward R, Harkness W. Reduced bacterial adhesion to hydrocephalus shunt catheters mediated by cerebrospinal fluid proteins. J Neurol Neurosurg Psychiatr. 1996;60:671-675.

[93] Fulkerson DH, Vachhrajani S, Bohnstedt BN, Patel NB, Patel AJ, Fox BD, Jea A, Boaz JC. Analysis of the risk of shunt failure or infection related to cerebrospinal fluid cell count, protein level, and glucose levels in low-birth-weight premature infants with posthemorrhagic hydrocephalus. J neurosurg Pediatr. 2011;7:147-151. DOI: 10.3171/2010.11.PEDS10244

[94] Simon TD, Kronman MP, Whitlock KB, Gove NE, Mayer-Hamblett N, Browd SR, Cochrane DD, Holubkov R, Kulkarni AV, Langley M, Limbrick DD, Luerssen TG, Oakes WJ, Riva - Cambrin J, Rozelle C, Shannon C, Tamber M, Wellons JC, Whitehead WE, Kestle JRW. Reinfection after treatment of first cerebrospinal fluid shunt infection: a prospective observational cohort study. J Neurosurg Pediatr. 2018;21:346-358. DOI: 10.3171/2017.9. PEDS17112

[95] Choux M, Genitori L, Lang D, Lena G. Shunt implantation: reducing the incidence of shunt infection. J Neurosurg. 1992;77:875-880.

[96] Choksey MS, Malik IA. Zero tolerance to shunt infections: can it be achieved? J Neurol Neurosurg Psychiatr. 2007;75:87-91.

[97] Kestle JRW, Holubkov R, Cochrane DD, Kulkarni AV, Limbrick DD, Luerssen TG, Oakes WJ, Riva-Cambrin J, Rozelle C, Simon TD, Walker ML, Wellons JC, Browd SR, Drake JM, Shannon CN, Tamber MS, Whitehead WE. A new Hydrocephalus Research Network protocol to reduce cerebrospinal fluid shunt infection. J Neurosurg Pediatr. 2016;17:391-396. DOI: 10.3171/2015.8.PEDS15253

[98] Pirotte BJ, Lubansu A, Bruneau M, Loqa C, Van Cutsem N, Brotchi J: Sterile surgical technique for shunt placement re- duces the shunt infection rate in children: preliminary analy- sis of a prospective protocol in 115 consecutive procedures. Childs Nerv Syst 2007;23:1251-1261.

[99] Hommelstad J, Madsø A, Eide PK: Significant reduction of shunt infection rate in children below 1 year of age after implementation of a perioperative protocol. Acta Neurochir (Wien). 2013;155:523-531.

[100] Hooper GJ, Rothwell AG, Frampton C, Wyatt MC. Does the use of

laminar flow and space suits reduce early deep infection after total hip and knee replacement? J Bone Joint Surg. 2011;93:85-90.

[101] Gastmeier P, Breier AC, Brandt C. Influence of laminar airflow on prosthetic joint infections: a systematic review. J Hosp Infect. 2012;81:73-78.

[102] Thompson DNP, Hartley JC, Hayward RD. Shunt infection: is there a near-miss scenario? J Neurosurg. 2007;106:15-19.

[103] Velghe L, Dereymaeker A, van der Voorde H. Swabbing of operative field in neurosurgery: analysis of 1000 controls. Acta Neurosurg (Wien). 1964; II:686-693.

[104] Tabara Z, Forrest DM. Colonisation of CSF shunts: preventive measures. Z Kinderchir. 1982;37:156-158.

[105] Fitzgerald R, Connelly B. An operative technique to reduce valve colonisation. Z Kinderchir. 1984;39(suppl II)107-109.

[106] Raahave D. Bacterial density in operation wounds. Acta Chir Scand.1974;8:585-593.

[107] Tulipan N, Cleves MA. Effect of an intraoperative double-gloving strategy on the incidence of cerebrospinal fluid shunt infection. J Neurosurg Pediatr. 2006;104:5-8.

[108] Selwyn S, Ellis H. Skin bacteria and skin disinfection reconsidered. Br Med J. 1972;1:136-140.

[109] Elek SO, Conen PE. The virulence of *Staphylococcus pyogenes* for man: a study of the problems of wound infection. Br J Exp Pathol. 1957;38: 573-86.

[110] Maiwald M, Chan ESY. The forgotten role of alcohol: a systematic review and meta-analysis of the clinical efficacy and perceived role of chlorhexidine in skin antisepsis. PLoS One. 2012;7: e44277. doi:10.1371/journal.pone.0044277

[111] Broekman MLD, van Beijnum J, Peul WC, Regli L. Neurosurgery and shaving: what's the evidence? J Neurosurg. 2011;115:670-678.

[112] Klimo P, Flannery AM. Pediatric hydrocephalus: systematic literature review and evidence – based guidelines Part 6: preoperative antibiotics for shunt surgery in children with hydrocephalus: a systematic review and meta-analysis. J Neurosurg Pediatr. 2015;16:237-239. DOI: 10.3171/2015.3.PEDS14326a

[113] Ragel BT, Brown SR, Schmidt RH. Surgical shunt infection: significant reduction when using intraventricular and systemic agents. J. Neurosurg. 2006;105:242-247.

[114] Rozelle CJ, Leonardo J, Li V. Antimicrobial suture wound closure for cerebrospinal fluid shunt surgery: a prospective, double-blinded, randomized controlled trial. J Neurosurg Pediatr. 2008;2:111-117. DOI: 10.3171/PED/2008/2/8/111

[115] Krause M, Mahr CV, Dchob S, Nestler U, Wachowiak R. Topical instillation of vancomycin lowers the rate of CSF shunt infections in children. Child's Nerv Syst. 2019;35:1155-1157. doi.org/10.1007/s00381-019-04185-1

[116] Bridgett MJ, Davies MC, Denyer SP, Eldridge PR. In vitro assessment of bacterial adhesion to hydromer – coated cerebrospinal fluid shunts. Biomaterials. 1993;14:184-188.

[117] Bayston R, Bhundia C, Ashraf W. Hydromer – coated catheters to prevent shunt infection? J Neurosurg Pediatr. 2005;102:207-212.

[118] Kaufmann AM, Lye T, Redekop G, Brevner A, Hamilton M, Kozey M,

Easton D. Infection rates in standard vs hydrogel coated ventricular catheters. Can J Neurol Sci. 2004;31:506-510.

[119] Kestle JRW, Riva-Cambrin J, Wellons JC, Kulkarni AV, Whitehead WE, Walker ML, Oakes WJ, Drake JM, Luersssen TG, Simon TD, Holubkov R. A standardized protocol to reduce cerebrospinal fluid shunt infection: the hydrocephalus clinical research network quality improvement initiative. J Neurosurg Pediatr. 2011;8:22-29. DOI: 10.3171/2011.4. PEDS10551

[120] Bayston R, Lambert E. Duration of protective activity of cerebrospinal fluid shunt catheters impregnated with antimicrobial agents to prevent bacterial catheter- related infection.J Neurosurg. 1997;87:247-251.

[121] Thomas R, Lee S, Patole S, Rao S. Antibiotic – impregnated catheters for the prevention of CSF shunt infections: a systematic review and meta-analysis. B J Neurosurg. 2012;26:175-184. DOI: 10.3109/02688697.2011.603856

[122] Edwards NC, Engelhart L, Casamento EM, McGirt MJ: Cost-consequence analysis of antibiotic-impregnated shunts and external ventricular drains in hydrocephalus. J Neurosurg. 2015;122:139-147.

[123] Lockhart PB, Loven B, Brennan MT, Fox PC. The evidence base for the efficacy of antibiotic prophylaxis in dental practice. J Amer Dent Assoc. 2007;138:458-474.

[124] Foreman PM, Hendrix P, Griessenauer CJ, Schmalz PG, Harrigan MR. External ventricular drain placement in the intensive care unit versus operating room: evaluation of complications and accuracy. Clin Neurol Neurosurg. 2015;128:94-100.

[125] Gigante P, Hwang BY, Appelboom G, Kellner CP, Kellner MA, Connolly ES. External ventricular drainage following aneurysmal subarachnoid haemorrhage. B J Neurosurg. 2010;24:625-632. DOI: 10.3109/02688697.2010.505989

[126] Flibotte JJ, Lee KE, Koroshetz WJ, Rosand J, McDonald CT. Continuous Antibiotic Prophylaxis and Cerebral Spinal Fluid Infection in Patients with Intracranial Pressure Monitors. Neurocrit Care. 2004;1:61-68.

[127] Wong GKC, Poon WS, Lyon D, Wai S. Cefepime vs. Ampicillin/Sulbactam and Aztreonam as antibiotic prophylaxis in neurosurgical patients with external ventricular drain: result of a prospective randomized controlled clinical trial. J Clin Pharm Therapeut. 2006;31:231-235.

[128] Wong GK, Ip M, Poon WS, Mak CW, Ng RY. Antibiotics-impregnated ventricular catheter versus systemic antibiotics for prevention of nosocomial CSF and non-CSF infections: a prospective randomised clinical trial. J Neurol Neurosurg Psychiatr. 2010;81:1064-1067.

[129] Dellit TH, Chan JD, Fulton C, et al. Reduction in *Clostridium difficile* infections among neurosurgical patients associated with discontinuation of antimicrobial prophylaxis for the duration of external ventricular drain placement. Infect Control Hosp Epidemiol. 2014;35:589-590.

[130] Alleyne CH, Hassan M, Zabramski JM. The efficacy and cost of prophylactic and perioprocedural antibiotics in patients with external ventricular drains. Neurosurg. 2000;47:1124-1127.

[131] Murphy RKJ, Liu B, Srinath A, Reynolds MR, Liu J, Craighead MC, Camins BC, Dhar R, Kummer TT, Zipfel GJ. No additional protection against ventriculitis with prolonged systemic antibiotic prophylaxis for

patients treated with antibiotic-coated external ventricular drains. J Neurosurg. 2015;122:1120-1126. DOI: 10.3171/2014.9.JNS132882

[132] Fichtner J, Guresir E, Seifert V, Raabe A. Efficacy of silver- bearing external ventricular drainage catheters: a retrospective analysis. J Neurosurg. 2010;112:840-846.

[133] Lemcke J, Depner F, Meier U. The impact of silver nanoparticle-coated and antibiotic-impregnated external ventricular drainage catheters on the risk of infections: a clinical comparison of 95 patients. Acta Neurochir Suppl. 2012;114:347-350.

[134] Lajcak M, Heideche V, Haude KH, Rainov NG. Infection rates of external ventricular drains are reduced by the use of silver-impregnated catheters. Acta Neurochir. 2010; 155:875-881. DOI 10.1007/s00701-013-1637-9

[135] Zakaria R, Tripathy S, Srikandarajah N, Rothburn MM, Lawson DD. Reduction of drain-associated cerebrospinal fluid infections in neurosurgical inpatients: a prospective study. J Hosp Infect. 2013;84:215-221.

[136] Lwin S, Low SW, Choy DKS, Yeo TT, Chou N. External ventricular drain infections: successful implementation of strategies to reduce infection rate. Singapore Med J. 2012;53:2555-259.

[137] Keong NC, Bulters DO, Richards HK, Farrington M, Sparrow OC, Pickard JD, Hutchinson PJ, Kirkpatrick PJ. The SILVER (Silver Impregnated Line Versus EVD Randomized trial): a double-blind, prospective, randomized, controlled trial of an intervention to reduce the rate of external ventricular drain infection. Neurosurg. 2012;71:394-403.

[138] Atkinson RA, Fikrey L, Vail A, Patel HC. Silver-impregnated external -ventricular -drain -related cerebrospinal fluid infections: a meta-analysis. J Hosp Infect. 2013;92:263-272. DOI.org/10.1016/j.jhin.2015.09.014

[139] Bayston R, Vera L, Mills A, Ashraf W, Stevenson O, Howdle SM. Antimicrobial activity of silver-processed catheters for neurosurgery. J Antimicrob Agents Chemother. 2010;65:258-265. doi:10.1093/jac/dkp420

[140] Zabramski JM, Whiting D, Darouiche RO, Horner TG, Olson J, Robertson C, Hamilton AJ. Efficacy of antimicrobial- impregnated external ventricular drain catheters: a prospective, randomized, controlled trial. J Neurosurg. 2003;98:725-730.

[141] Abla AA, Zabramski JM, Jahnke HK, Fusco D, Nakaji P: Comparison of two antibiotic-impregnated ventricular catheters: a prospective sequential series trial. Neurosurg. 2011; 68:437-442.

[142] Muttaiyah S, Ritchie S, John S, Mee E, Roberts S. Efficacy of antibiotic-impregnated external ventricular drain catheters. J Clin Neurosci. 2010;17:296-298. doi:10.1016/j.jocn.2009.06.016

[143] Tamburrini G, Massimi L, Caldarelli M, Di Rocco C. Antibiotic impregnated external ventricular drainage and third ventriculostomy in the management of hydrocephalus asso- ciated with posterior cranial fossa tumours. Acta Neurochir (Wien) 2008;150:1049-1055.

[144] Bayston R, Ashraf W, Pelegrin I, Fowkes K, Bienemann AS, Singleton WGB, Scott IS. An external ventricular drainage catheter impregnated with rifampicin, trimethoprim and triclosan, with extended activity against MDR gram negative bacteria: an invitro and in vivo

study. J Antimicrob Chemother. 2019;74:2959-1264. doi:10.1093/jac/dkz293

[145] Khanna RK, Rosenblum ML, Rock JP, Malik GM (1995) Prolonged external ventricular drainage with percutaneous long-tunnel ventriculostomies. J Neurosurg. 1995;83:791-794. DOI 10.1007/s00701-007-1458-9

[146] Collins CDE, Hartley JC, Chakraborty A, Nolan D, Thompson P. Long subcutaneous tunnelling reduces infection rates in paediatric external ventricular drains. Child's Nerv Syst. 2014;30:1671-1678. DOI: 10.1007/s00381-014-2523-3

[147] Leung GKK, Ng KB, Taw BBT, Fan YW. Extended subcutaneous tunnelling technique for external ventricular drainage. B J Neurosurg. 2007;21:359-364. DOI: 10.1080/02688690701392881

[148] Mayhall CG, Archer NH, Lamb VA, Spadora AC, Baggett JW, Ward JD, Narayan RK. Ventriculostomy-related infections. A prospective epidemiologic study. N Engl J Med. 1984;310:553-559.

[149] Wong GK, Poon WS, Wai S, Yu LM, Lyon D, Lam JM. Failure of regular external ventricular drain exchange to reduce cerebrospinal fluid infection: result of a randomised controlled trial. J Neurol Neurosurg Psychiatr.2002; 73:759-761.

[150] Lo CH, Spelman D, Bailey M, Cooper DJ, Rosenfeld JV, Brecknell JE. External ventricular drain infections are independent of drain duration: an argument against elective revision. J Neurosurg. 2007;106:378-383.

[151] Mayer C, Albert R, Proescholdt MA, Bele S, Woertgen C, Brawanski A. Can a regular change of external ventricular drainage (EVD) prevent cerebrospinal fluid infection in patients with intracranial hemorrhage? German Med Sci. 2006; www.egms.de/en/meetings/dgnc2006/06dgnc250.shtml

[152] Bianco A, Quirino A, Giordano M, Marano V, Rizzo C, Liberto MC, Foca A, Pavia M. Control of carbapenem-resistant Acinetobacter baumannii outbreak in an intensive care unit of a teaching hospital in southern Italy. BMC Infect Dis. 2016;16:747. DOI: 10.1186/s12879-016-2036-7

[153] Chia PY, Sengupta S, Kukreja A, Ponnampalavanar SSL, Ng OT, Marimutjhu K. The role of hospital environment in transmission of multidrug-resistant gram -negative organisms. Antimicrob Res Infect Control. 2020;9:1-11. DOI.org/10.1186/s13756-202-0685-1

[154] Moon HJ, Kim SD, Lee JB, Lim DJ, Park JY. Clinical analysis of external ventricular drainage related ventriculitis. J Korean Neurosurg Soc. 2007;41:236-240.

[155] Leverstein-van Hall MA, Hopmans TE, van der Sprenkel JW, Blok HE, van der Mark WA, Hanlo PW, Bonten MJM. A bundle approach to reduce the incidence of external ventricular and lumbar drain-related infections. J Neurosurg. 2010;112:345-353. DOI: 10.3171/2009.6.JNS09223

[156] Korinek A-M, Reina M, Boch AL, Rivera AO, DE Bels D, Puybasset L. Prevention of external ventricular drain-related ventriculitis. Acta Neurochir (Wien) 2005;147:39 – 45. DOI 10.1007/s00701-004-0416-z

[157] Dasic D, Hanna SJ, Bojanic S, Kerr RSC. External ventricular drain infection: the effect of a strict protocol on infection rates and a review of the literature. Br J Neurosurg 2006;20:296-300. DOI: 10.1080/0268869060 0999901

Section 5

Germinal Matrix Hemorrhage

Chapter 7

Germinal Matrix-Intraventricular Hemorrhage: Current Concepts and Future Direction

Sadhika Sood and Rohit Gulati

Abstract

Germinal Matrix Hemorrhage-Intraventricular hemorrhage (IVH) is a bleed of multifactorial etiology involving the highly vascular and delicate neuro-glial precursors in the developing brain. It poses a challenging complication in preterm newborns. This chapter provides a focused discussion on the current concepts in pathogenesis, management, and complications of IVH. The radiological findings at diagnosis and follow-up and the cytological features of CSF will be valuable to both frontline and diagnostic healthcare providers. The chapter also reviews the ongoing scientific development in the field. The authors believe that this chapter will be a valuable tool for all healthcare providers (students, physicians, and in nursing care) in managing this challenging condition.

Keywords: Germinal matrix hemorrhage, intraventricular hemorrhage, IVH, intracranial hemorrhage, superficial siderosis, central nervous system, cerebrospinal fluid, genetic alterations, cranial ultrasound, preterm complications, low birth weight

1. Introduction

The germinal matrix (GM) is a specialized layer of glial and neuronal precursor cells in the periventricular region of the brain with high metabolic activity, which is strongly dependent on its rich vascularity and rapid angiogenesis [1]. The dense and fragile vasculature makes GM selectively vulnerable to hemorrhage. Germinal matrix – intraventricular hemorrhage (GM-IVH) is the most common type of intracranial hemorrhage in preterm infants. A combination of increased perinatal stress, poor cerebral autoregulation, and inherent fragility of the nascent vessels in the germinal matrix increases the likelihood of the development of GM-IVH in preterm infants. Also, there is evidence of occurrence in-utero and among full-term infants, however, such cases are rare [2]. The germinal matrix disappears by 36–37 weeks of gestation (wg), so GM-IVH is more likely in preterm infants than full term.

The global incidence of GM-IVH among preterm infants ranges from 14.7% to 44.7%, with variations across gestational age groups, countries, and antenatal and neonatal care [3]. The widespread use of cranial ultrasonography since the early 1980s, increasing knowledge of risk factors, antenatal steroid usage, and improved intensive care have improved incidence, survival, and morbidity of GMH [4]. However, GMH continues to remain a significant healthcare issue in preterm infants and a recognizable cause of long-term neurological and behavioral issues in survivors.

2. Germinal matrix-intraventricular hemorrhage

2.1 Pathogenesis

Developmentally, GM is located in the ganglionic eminence of the brain and is most pronounced in the caudate nucleus. The thickness and density of GM vasculature are higher than other brain areas and begin to decrease after 24 weeks of gestation (wg) and almost disappear at 36–37 wg with increasing fetal maturity [1, 5]. A significant bleed in the highly vascular GM breaks the associated ependyma to involve the lateral cerebral ventricle constituting intraventricular hemorrhage (IVH) [6, 7]. The incidence of GMH-IVH increases with decreasing gestation age at birth in preterm infants [8–10].

The pathogenesis of GM-IVH is complex and heterogeneous. The blood–brain barrier (BBB) associated with GM vasculature is distinct from the remaining areas in the brain due to diminished: 1) pericytes, 2) fibronectin in the basal lamina, and 3) GFAP (glial fibrillary acidic protein) in astrocyte endfeet (**Figure 1**). The paucity of three essential components of the BBB leads to the altered structural integrity of GM vasculature. First, pericytes play an essential role in BBB development, especially in early angiogenesis, extracellular matrix production, and endothelial maturation [11]. The paucity of pericytes in GM is associated with diminished levels of TGF-β [12] and predisposition to hemorrhage in dilatated blood vessels in experimental models [13]. Second, fibronectin, a high molecular weight glycoprotein, is selectively deficient in the GM basement membrane [14]. Fibronectin polymerizes to provide structural integrity to blood vessels and is dependent on TGF-β for its upregulation. While other basement membrane components, including Collagen I, II, IV, laminin, and perlecan, are similar to other components in the human brain [14, 15]. Third, astrocytes provide vascular integrity by sheathing the predominance of the BBB with their GFAP rich extensions (endfeet). Autopsy studies in premature infants show decreased GFAP expressing astrocyte endfeet in GM than

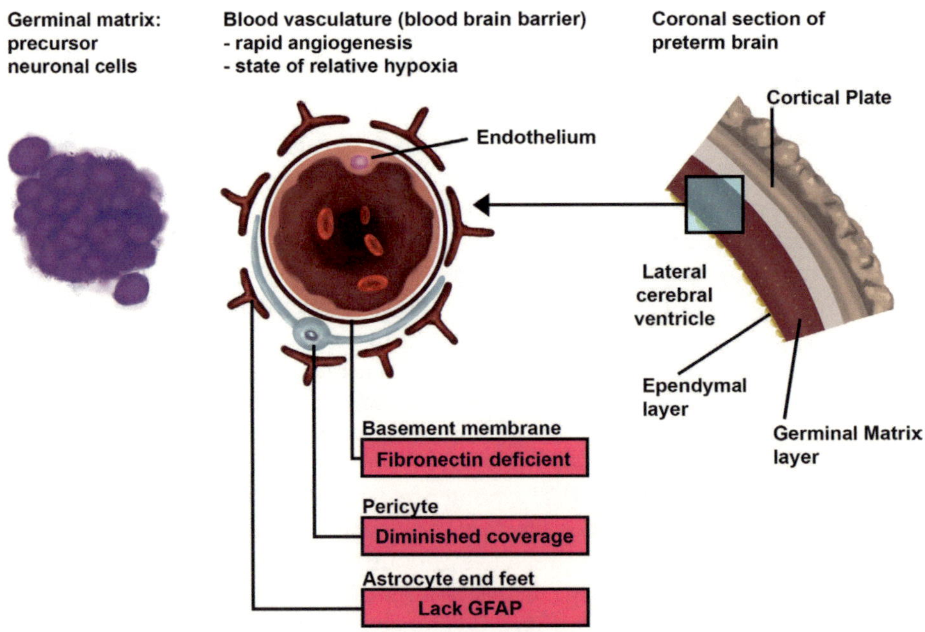

Figure 1.
Diagrammatic representation of the coronal section of a preterm brain to highlight the factors contributing to the labile structure of the blood brain barrier in the germinal matrix and pathogenesis of the GM-IVH.

cerebral cortex and white matter [16]. These make the blood–brain barrier fragile and more susceptible to hemorrhage.

Microscopically the GM vasculature has been described as circular in coronal sections, compared to elongated and flat vessels in other areas of the brain, representing the immaturity of the vessels from rapid angiogenesis and high endothelial turnover [17]. In addition, immunofluorescence and electron microscopy have shown a paucity of pericytes in the GM vascular environment [12].

Finally, fluctuations in cerebral blood flow precipitate into hemorrhage in the delicate GM. In addition, defects in the hemostatic mechanisms expectantly promote hemorrhage [6, 7].

Germinal matrix cells being metabolically active precursor neuronal and glial cells in the early stages of maturation demand a specialized and rich blood supply. This requirement is met by accelerated angiogenesis dependent on high levels of vascular endothelial growth factors (VEGF) and angiopoietin-2 and low expression of TGF-β [1]. Also, the GM is in a state of relative hypoxia, a driving force for continuous angiogenesis [6, 7]. Intriguingly, this may explain the near absence of GM-IVH after over 3–5 days of birth irrespective of the duration of gestation. Likely, higher oxygenation following birth inhibits rapid angiogenesis. Thus, a labile combination of metabolically active immature/precursor cells with a rich but "structurally weak" vasculature provided a high-risk background for bleeding, especially with high-velocity cerebral blood flow.

Among many factors associated with alteration in cerebral blood flow, severe respiratory distress syndrome, patent ductus arteriosus, high central venous, and hypercarbia are most prominent. While autoregulation maintains constant cerebral blood flow, this mechanism is impaired in premature infants with lower birth weight. Thus, changes in blood volume or pressure are more likely to affect cerebral circulation. Interestingly, the results of studies directly comparing impaired autoregulation with GMH-IVH have been mixed [18–20] and provide an opportunity for further research in this direction. As seen in pneumothorax and mechanical ventilation (on high mean airway pressure mode), high central venous pressure stands out as a solid contender to contribute to IVH. This is also concordant with the venous nature of GM-IVH [21]. Interestingly, mechanical ventilation in synchronized and intermittent mandatory mode prevents higher velocity/turbulence of cerebral blood than fixed frequency/pressure modes.

Significant other risk factors affecting include prolonged labor, maternal chorioamnionitis, early-onset sepsis, development of respiratory distress, recurrent tracheal suctioning (supportive care especially during mechanical ventilation), and hypoxia. While most of these factors impact cerebral blood flow, infectious and hypoxic etiologies alter the GM microvasculature. The role of hypotension and rapid sodium bicarbonate infusion in the causation of IVH are inconclusive.

2.2 Diagnosis

Clinical manifestations of GM-IVH include asymptomatic to subtle alterations in consciousness, limb and eye movement, and changes in muscular tone following IVH. Further, severe cases may be associated with cardiorespiratory distress and progression to seizures, hypotonia, or decerebrate posturing [22].

Cranial ultrasound (CUS) remains the most practical and well-utilized approach for diagnosing and monitoring GM-IVH evolution. Newer ultrasound devices with high-frequency transducers allow for enhanced evaluation. Epidemiologically, surviving infants born preterm at 24 weeks have a higher incidence (10–25%) of high-grade GM-IVH (grade 3–4) as compared to preterm infants born after 28 weeks (<5%) [8–10]. Almost half the cases of postnatal GM-IVH present on the first day of life, with nearly ~90% presenting within the first 72 hours. As discussed in pathogenesis, increased oxygenation after birth likely stabilizes the GM-BBB and makes infants almost resistant to GM-IVH after the first week of life irrespective of gestational age [23]. Therefore, regular CUS schedules have been recommended based on the gestational age at birth and when otherwise clinically indicated [24].

Traditionally, GM-IVH had been graded into four categories based on the extent of hemorrhage beginning in the venule that drains into the subependymal collector veins: grade-1 representing subependymal hemorrhage; grade-2 with limited (filling <50% of normal-sized ventricles) IVH; and grade-3 with extensive IVH. Grade-4 was defined as IVH with parenchymal extension [25]. However, the latter was better identified as parenchymal venous infarction (PVI), though parenchymal extension does also rarely occurs [26]. Interestingly, PVI may occur in all, including lower grades (1 and 2) of GM-IVH [22]. Since PVI is associated with long-term complications and risk of mortality (based on location and extent), a three-stage grading with an additional description of PVI has been recommended [22, 24] (**Figure 2**). In addition, early GMH may alter local neuronal and glial precursors with neurological consequences, description of location of bleed in addition to grade is suggested.

On CUS, grade 1 GMH is subependymal, hyperechoic, and globular. Evaluation in both coronal and sagittal planes helps distinguish a small GMH from choroid plexus on an initial diagnostic scan. Also, echogenicity at the caudothalamic groove (usual site for GMH) in the late neonatal period likely represents hyperechoic germinolysis and not late GMH [27]. Distinguishing pure subependymal bleed from IVH may be challenging on CUS. Indirect signs of hyperechoic ependymal changes, which usually occur 2 to 4 weeks after IVH, and insonation through mastoid fontanelle are helpful in this distinction [24] and aid prognostication and counseling.

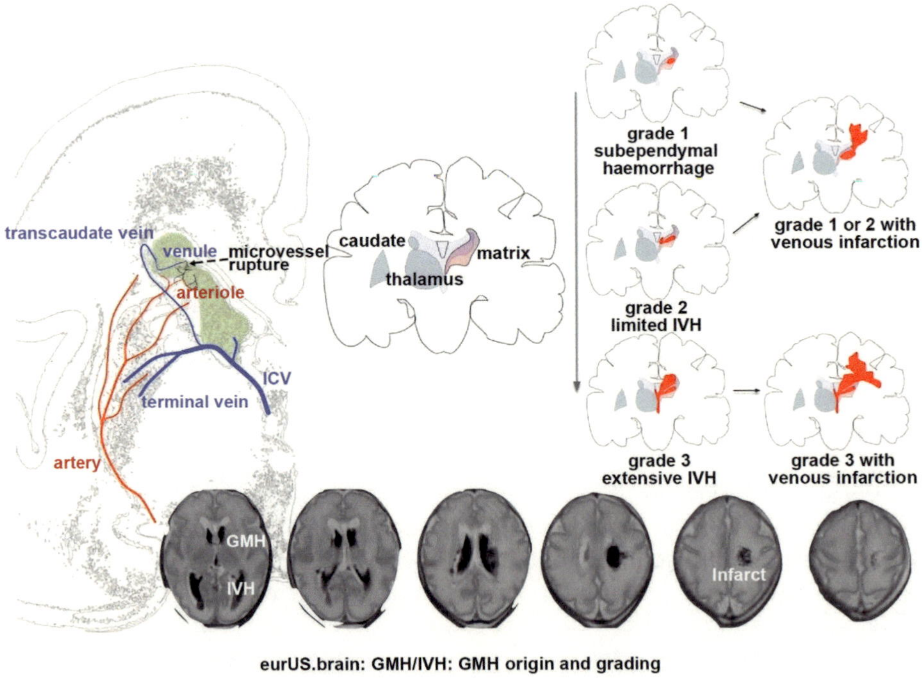

Figure 2.
GMH/IVH: Origin and grading. GMH starts in a venule that drains into lateral subependymal collector veins; it extends into white matter by virtue of venous compression and infarction; bottom row: T2-weighted MRI of GMH with limited IVH and limited venous infarct. (Derived from Parodi et al. [24]).

Clot changes overtime should also be recorded. A subacute clot or clot remnants early after birth may represent an antenatal hemorrhage.

PVI typically is identified as a triangular echo density in the periventricular white matter adjacent to the GMH. The infarct may not touch the GMH initially and may or may extend into the GMH depending on severity. Infarcts eventually evolve into cavitary lesions, and porencephaly ensues in 1–2 months [24]. This cavitation is asymmetric, unilateral, and permanent in contrast to cysts of periventricular leukomalacia (symmetric, bilateral, and transient) [22].

A quarter of infants with GM-IVH develop posthemorrhagic ventricular dilatation (PHVD) due to imbalanced production and resorption of CSF. This dilatation occurs a few days to weeks after IVH and is followed by subsequent regression [28]. PHVD is more common in higher grades but can occur in all cases with IVH. Thus, serial CUS is recommended in IVH cases until term-equivalent age. While a subset of cases resolves spontaneously, balancing the complications of compression versus those of surgical management (tapping, shunt) remains a challenge [29]. PVHD, as expected, is associated with a poor neurological outcome in the long term.

2.3 Genetic factors in GM-IVH

Thrombophilic genotype is frequently associated with a subset of severe GM-IVH patients with atypical clinical presentation. The atypical presentation includes periventricular hemorrhagic infarction presenting within 6 hours of birth or after four days of birth, in the absence of secondary inciting factors like sepsis. Factor V Leiden mutation was the most common genetic alteration, frequently with mothers being carriers. Prothrombin mutations and polymorphism of the *MTHFR* gene were also reported [30]. Previous studies have shown an association of a thrombophilic profile with early grade 1–2 GMH [31, 32]. Polymorphism of TNF-α has been associated with an increased risk of GMH-IVH [33]. Of interest, the same study showed an association of polymorphism in TGF-β with a fatal outcome but not with IVH.

Mutations of the *COL4A1* gene, coding for type IV collagen *α*-chain-1, have rarely been reported in a subset of preterm IVH [30, 34]. A pair of dizygotic twins showed a heterozygous duplication at exon 4 of the highly conserved and ubiquitous *COL4A1.* Interestingly, the mother and maternal grandmother of the twins were heterozygous carriers and were asymptomatic. Also, the study evaluated 39 other cases of preterm IVH and detected no mutations in *COL4A1,* indicating its rarity [34]. Previous studies have revealed no alteration in type IV collagen components in the basement membrane in GM [14, 15].

Experimental models have shown tropomyosin receptor kinase B (TrkB) to influence the inflammatory status in the microenvironment following GMH by influencing the phosphatidylinositol-3-kinases (PI3K)/protein kinase B (Akt)/forkhead box protein O1 (FoxO1) pathway [35].

Overall, genetic alteration in components of vascular structure, coagulation mechanism, and inflammatory pathways have been described in a subset of GM-IVH. The authors believe that recent progress in inflammation and growing knowledge of inflammasome complex may be employed towards further research in this direction.

2.4 Prevention

Our understanding of IVH due to a structurally labile and immature vasculature in the germinal matrix and alterations in cerebral blood flow in premature infants forms the focus of most strategies to prevent GM-IVH. In principle, delay of preterm birth relies on decreasing GM vascular density with advanced gestational age. Moreover, high postnatal oxygen levels in the infant mediate the stabilization of the GM blood vasculature and ensure freedom from IVH in 3–5 days after birth, highlighting the critical importance of timeliness in management and prevention.

2.4.1 Specific strategies for prevention

Steroids (glucocorticoids) like dexamethasone and betamethasone are administered to pregnant women in premature labor under 34 wg. Glucocorticoids cause a selective inhibition of blood vessels in the GM- BBB that lack adequate pericyte coverage, inhibit angiogenesis, and subsequently stabilize vasculature [12, 14, 36]. In addition, prenatal corticosteroid assists in development lungs surfactant and protect against respiratory distress syndrome. The latter effect also prevents turbulent cerebral blood flow. Prenatal corticosteroid usage is one of the rare factors that has consistently been associated with a reduction in occurrence and severity of IVH [37, 38].

Indomethacin is a non-selective cyclooxygenase (COX) inhibitor and reduces severe IVH, especially in males [39, 40]. Indomethacin is employed for closure of patent ductus arteriosus that in turn prevents altered cerebral blood flow. It also suppresses angiogenesis by COX-2 inhibition [1]. Although indomethacin can decrease IVH in the short term, its usage is not associated with reducing long-term neurological complications such as cerebral palsy, deafness, and blindness [41–43]. Hence, indomethacin has limited acceptance and is based on regional preferences.

Prenatal care and transport: It is recommended that pregnant mothers be given adequate antenatal care and those in preterm labor be transported (while pregnant) to tertiary care units better equipped to manage both mother and child. Transportation of extremely premature infants has long been associated with the increased occurrence and severe IVH [44].

It is beneficial to note that antenatal phenobarbital and magnesium, vitamin-K, and fresh frozen plasma did not influence the occurrence of IVH [45–49].

Intriguing preclinical studies show time-sensitive windows for therapeutic pharmacological targeting of the GM “weakened” BBB by altering the integrin-β8 and TGF-β pathways [50].

2.5 Management

Currently, there is a paucity of active treatment strategies for the management of established GM-IVH. Maintaining blood pressure levels and respiratory status, with judicious use of IV fluids, blood transfusions, and respiratory support (if needed), might prevent the progression of hemorrhage. Electroencephalogram (EEG) monitoring should be done in the presence of seizures [3]. Apart from supportive treatment, emphasis is laid on the preservation of cerebral perfusion and the prevention of complications. Monitoring twice weekly with CUS for four weeks (or similar) and then weekly till term equivalent age recommended to evaluate GMH and post hemorrhage hydrocephalus (PHH).

2.5.1 Post-natal effective nursing care

Multiple trials and observational studies have focused on the relative head position of premature infants soon after birth in relation to IVH. These positional strategies focus mainly on maintaining adequate cerebral blood flow.

While previous studies on the effect on neutral head position found no significant association with the occurrence of IVH [51], these studies were also limited by small sample size [52]. More recently, efficient, supportive nursing intervention in premature infants during the first 72 hours of birth has been associated with decreased incidence and progression of GM-IVH [53]. This four-pronged approach includes midline head position, head elevation of the incubator, and slow vascular flushing/withdrawal of blood, and sudden elevation of the legs. First, the head in midline position ensures adequate venous drainage. Head rotation impedes jugular venous outflow on the ipsilateral side and may cause congestion, relative hypoxia and eventually aid GMH [54]. Second, incubator head lift (15–30 degrees) enhances gravitational cerebral venous drainage [55]. Third, sudden elevation of legs, as in to change diapers, may result in increased venous return, increase cardiac preload, thereby altering cerebral perfusion. Finally, avoiding rapid (lasting <30 seconds) vascular flushing/blood collection can avoid a transient though significant alteration in cerebral blood flow [56]. The effect of the intervention was stronger in infants born before 27 wg [53]. While previous studies on the effect on neutral head position found no significant association with the occurrence of IVH [51], these studies were limited by small sample size [52]. A more recent meta-analysis showed the limited utility of supine midline head position for the prevention of GM-IVH. However, midline head position with an elevation of incubator head was associated with lower mortality [57]. Overall, concomitant intervention with neutral head position, the elevation of incubator head, and avoidance of sudden leg elevation and sudden vascular volumetric changes provide evidence for a better outcome.

2.6 Complications and potential treatment strategies

The survivors of severe GMH frequently develop post-hemorrhagic hydrocephalus (PHH). A subset of these cases requires surgical shunting, which is not without its complications, including infections, obstruction, and displacement [58]. In addition, the cerebroventricular dilatation causes physical pressure on the brain parenchyma and is associated with neurological impairment in the long term. Mechanism of PHH: Obstruction of the cerebral aqueduct, foramina of Luschka and Magendie, and subarachnoid outflow passages by blood clots/microthrombi may cause PHH. Historically, fibrinolytic therapy has not been successful in the management of PHH.

The tissue macrophage system responds to intracranial hemorrhage similar to other locations in the body. Red blood cells (RBCs) are phagocytosed by macrophages (erythrophages), and subsequently, hemoglobin is degraded. Iron mainly converts to coarse, irregular hemosiderin granules and porphyrin rings into bilirubin. In exceptional circumstances with closed compartments and lower oxygen tension, such as intracranial bleed, hematoidin, a crystalline, reduced biliverdin product may be formed. Post hemorrhagic components are frequently encountered on light microscopic evaluation of the cerebrospinal fluid (CSF), as early as 1–2 days after bleeding [59, 60]. In addition to the erythrophagocytosis, cellular components of the ventricular lining (ependymal cells and choroid plexus cells) and rarely, precursor germinal matrix cells (due to close proximity with disrupted ventricular lining) may be identified in CSF analysis [61–63].

Superficial siderosis (SS) is the deposition of hemosiderin in the subpial layers of CNS, resulting in sensorineural hearing loss and cerebellar ataxia in most adults cases [64]. Susceptibility weighted imaging (SWI), an MRI sequence, identified SS in the ependymal layer, brain stem, cerebellum with vermis, and Sylvian fissures. Interestingly the depth of SS correlated with the increasing grade of GM-IVH. Also, brain stem and cerebellar SS appear to relate more to IVH than cerebellar hemorrhage [65].

A review of scientific literature shows the following current trends exploring the management of GMH and prevention of complications.

2.6.1 Iron in the manifestation of PHH

Experimental models have shown the role of iron (from red blood cells) to develop brain edema and acute ventricular dilatation [66]. As proof of principle, iron chelation with deferoxamine has showed reduced long term PHH after GMH in neonatal rats [67, 68]. Another group found biliverdin reductase to enhance CD36 expression in scavenging microglia and hematoma resolution through NOS/TLR4 pathway [69]. Additionally, iron overload has been associated with increased aquaporin-4 expression [70]. However, diuretic treatment has not been found to be beneficial.

Along similar lines, "normal appearing" white matter in preterm infants with severe GM-IVH, at term equivalent age, showed paramagnetic (positive magnetic) susceptibility, likely due to diffusion of iron into the periventricular white matter [71]. This radiological finding may be employed as an innovative methodology for future research focusing on the spatial impact of iron deposition on long-term neurological consequences.

2.6.2 Role of inflammation and gliosis

Post GMH levels of pro-inflammatory markers like TNFα are elevated. In response to hemorrhage and associated tissue injury, resident microglia are activated in an inflammatory process [72–74]. Additional experimental models have shown microglial proliferation surrounding the clot with phosphorylated ERK. Minocycline and cannabinoid receptor-2 agonists have also shown promise to curb down inflammation [75]. CD200Fc inhibits inflammation following GMH likely by mediating CD200R1/Dok1 pathway [76]. IVH has been shown to cause a TLR4 and NF-κβ based inflammatory pathway mediated increase in CSF production in the choroid plexus. As a proof of principle, amelioration of these mediators was associated with control of CSF production and improvement in PHH [77]. The role of M2 microglia stimulation through the PPARγ and CD36 scavenger receptor for short-term resolution of hematoma has also shown promising results for further clinical evaluation [78]. NT-4 controls neuroinflammation by interacting with TrkB to induces PI3K-Akt pathway and inhibits downstream FoxO1 in experimental models [35]. These results promise potential for clinical utility in the management of PHH.

Extracellular matrix (ECM), especially components fibronectin and vitronectin, are elevated post-GMH and are hypothesized to deposit (like microthrombi), potentially causing CSF obstruction [75, 79–81]. TGF-β may be induced by thrombin and promotes the production of ECM, especially TGF-β1 isoform whose levels have been elevated in studies after GMH. It's inhibition has been associated with attenuated PHH and neurological decline [75, 81]. While GFAP expression is markedly increased in experimental IVH models, umbilical cord mesenchymal stem cell infusion has been associated with a decline in GFAP expression and subsequent PHH development [82]. The role of GFAP and astrocytes in gliosis post IVH requires further attention. More recently, astrogliosis was associated with redistribution of aquaporin-4 and altered CSF dynamics. Olomoucine controlled scarring and attenuated PHH by inhibition of cyclin-dependent kinase (CDK) [83]. Secukinumab, monoclonal IgG1κ targeting IL17a, is protective against reactive astrogliosis following GMH, partly by regulating IL-17RA/(C/EBPβ)/SIRT1 pathways [84].

2.6.3 Long term complications of low grade (grade 1: 2) GM-IVH

While neurological complications in survivors of high-grade GM-IVH are well documented, the impact of low-grade IVH currently continues to be better understood. Low-grade IVH was associated with moderate to severe neurodevelopmental impairment (NDI) and without association with cerebral palsy [85]. A case-controlled retrospective study using CUS found no significant impact of low-grade GM-IVH on neurological complications of cerebral palsy and neurodevelopmental delay evaluated during 18–30 months after birth [86]. Both these studies were limited in power and in analysis by more sensitive MR-based techniques [87]. A more recent MR-based study has revealed microstructural impairment of white matter related to neurodevelopmental impairment at 24 months in early GMH [88]. Similarly, magnetic resonance with 3D pseudo-continuous arterial spin-labeling (pCASL) perfusion sequence-based study has shown consistently lower CBF in the posterior cortical and subcortical gray matter regions in preterm neonates with low grade IVH [89]. This regional susceptibility also requires correlation with long term studies. From a developmental perspective, neurological alterations are not incompatible with low-grade IVH. GMH may lead to altered myelination in the white matter since ganglionic eminence is the seat of oligodendroglial precursor cells that migrate to cerebral white matter areas to produce myelin later in the third trimester [90]. Besides, GM is involved in the development of GABAergic interneurons significant for high-level cognitive function [91].

3. Conclusion

Germinal matrix intraventricular hemorrhage is the most common intracranial hemorrhage in newborns, particularly preterm neonates. Improvements in obstetric and neonatal care have led to increased survival of preterm infants. Despite extensive research and preventive measures, the incidence of associated complications and mortality remains high. The GM is highly susceptible to hemorrhage due to a combination of delicate vasculature and fluctuations of cerebral perfusion, uncontrolled by autoregulatory mechanisms. Genetic factors and coagulation disorders may factor in if present. Obstetric and neonatal clinicians should use the available knowledge to prevent the occurrence of and progressions of hemorrhages. Therapeutic options for the management of GM-IVH are predominantly limited to supportive care and monitoring. Shunts have proven to be effective in challenging cases of PHH. Current and ongoing improvement in the molecular understanding of GM-IVH and its complications using multi-omics investigations is essential to develop biomarkers and therapeutic strategies.

Acknowledgements

The authors thank Mr. Fredrik Skarstedt for his immense support with digital image preparation.

Conflict of interest

The authors declare no conflict of interest.

Author details

Sadhika Sood[1] and Rohit Gulati[2]*

1 Kasturba Medical College, Manipal Academy of Higher Education, Mangalore, India

2 Clinical Pathology, Indiana University School of Medicine, Indianapolis, Indiana, USA

*Address all correspondence to: rogulati@iu.edu

References

[1] Ballabh, P., et al., *Angiogenic inhibition reduces germinal matrix hemorrhage.* Nat Med, 2007. **13**(4): p. 477-485.

[2] Morioka, T., et al., *Fetal germinal matrix and intraventricular hemorrhage.* Pediatr Neurosurg, 2006. **42**(6): p. 354-361.

[3] Egesa, W.I., et al., *Germinal Matrix-Intraventricular Hemorrhage: A Tale of Preterm Infants.* Int J Pediatr, 2021. **2021**: p. 6622598.

[4] Yeo, K.T., et al., *Improving incidence trends of severe intraventricular haemorrhages in preterm infants <32 weeks gestation: a cohort study.* Arch Dis Child Fetal Neonatal Ed, 2020. **105**(2): p. 145-150.

[5] Ballabh, P., A. Braun, and M. Nedergaard, *The blood-brain barrier: an overview: structure, regulation, and clinical implications.* Neurobiol Dis, 2004. **16**(1): p. 1-13.

[6] Ballabh, P., *Intraventricular hemorrhage in premature infants: mechanism of disease.* Pediatr Res, 2010. **67**(1): p. 1-8.

[7] Ballabh, P., *Pathogenesis and prevention of intraventricular hemorrhage.* Clin Perinatol, 2014. **41**(1): p. 47-67.

[8] Fellman, V., et al., *One-year survival of extremely preterm infants after active perinatal care in Sweden.* Jama, 2009. **301**(21): p. 2225-2233.

[9] Ancel, P.Y., et al., *Survival and morbidity of preterm children born at 22 through 34 weeks' gestation in France in 2011: results of the EPIPAGE-2 cohort study.* JAMA Pediatr, 2015. **169**(3): p. 230-238.

[10] Stoll, B.J., et al., *Neonatal outcomes of extremely preterm infants from the NICHD Neonatal Research Network.* Pediatrics, 2010. **126**(3): p. 443-456.

[11] Armulik, A., et al., *Pericytes regulate the blood-brain barrier.* Nature, 2010. **468**(7323): p. 557-561.

[12] Braun, A., et al., *Paucity of pericytes in germinal matrix vasculature of premature infants.* J Neurosci, 2007. **27**(44): p. 12012-12024.

[13] Lindahl, P., et al., *Pericyte loss and microaneurysm formation in PDGF-B-deficient mice.* Science, 1997. **277**(5323): p. 242-245.

[14] Xu, H., et al., *Maturational changes in laminin, fibronectin, collagen IV, and perlecan in germinal matrix, cortex, and white matter and effect of betamethasone.* J Neurosci Res, 2008. **86**(7): p. 1482-1500.

[15] Anstrom, J.A., et al., *Morphometric assessment of collagen accumulation in germinal matrix vessels of premature human neonates.* Neuropathol Appl Neurobiol, 2005. **31**(2): p. 181-190.

[16] El-Khoury, N., et al., *Astrocyte end-feet in germinal matrix, cerebral cortex, and white matter in developing infants.* Pediatr Res, 2006. **59**(5): p. 673-679.

[17] Ballabh, P., A. Braun, and M. Nedergaard, *Anatomic analysis of blood vessels in germinal matrix, cerebral cortex, and white matter in developing infants.* Pediatr Res, 2004. **56**(1): p. 117-124.

[18] Tsuji, M., et al., *Cerebral intravascular oxygenation correlates with mean arterial pressure in critically ill premature infants.* Pediatrics, 2000. **106**(4): p. 625-632.

[19] Soul, J.S., et al., *Fluctuating pressure-passivity is common in the cerebral circulation of sick premature infants.* Pediatr Res, 2007. **61**(4): p. 467-473.

[20] Wong, F.Y., et al., *Impaired autoregulation in preterm infants identified by using spatially resolved spectroscopy.* Pediatrics, 2008. **121**(3): p. e604-e611.

[21] Ghazi-Birry, H.S., et al., *Human germinal matrix: venous origin of hemorrhage and vascular characteristics.* AJNR Am J Neuroradiol, 1997. **18**(2): p. 219-229.

[22] Volpe, J., *Neurology of the Newborn.* 2008, Philadelphia, PA: Saunders Elsevier. 1120.

[23] Dolfin, T., et al., *Incidence, severity, and timing of subependymal and intraventricular hemorrhages in preterm infants born in a perinatal unit as detected by serial real-time ultrasound.* Pediatrics, 1983. **71**(4): p. 541-546.

[24] Parodi, A., et al., *Cranial ultrasound findings in preterm germinal matrix haemorrhage, sequelae and outcome.* Pediatr Res, 2020. **87**(Suppl 1): p. 13-24.

[25] Papile, L.A., et al., *Incidence and evolution of subependymal and intraventricular hemorrhage: a study of infants with birth weights less than 1,500 gm.* J Pediatr, 1978. **92**(4): p. 529-534.

[26] Volpe, J.J., *Brain injury in the premature infant: overview of clinical aspects, neuropathology, and pathogenesis.* Semin Pediatr Neurol, 1998. **5**(3): p. 135-151.

[27] Horsch, S., P. Kutz, and C. Roll, *Late germinal matrix hemorrhage-like lesions in very preterm infants.* J Child Neurol, 2010. **25**(7): p. 809-814.

[28] Murphy, B.P., et al., *Posthaemorrhagic ventricular dilatation in the premature infant: natural history and predictors of outcome.* Arch Dis Child Fetal Neonatal Ed, 2002. **87**(1): p. F37-F41.

[29] Robinson, S., *Neonatal posthemorrhagic hydrocephalus from prematurity: pathophysiology and current treatment concepts.* J Neurosurg Pediatr, 2012. **9**(3): p. 242-258.

[30] Harteman, J.C., et al., *Atypical timing and presentation of periventricular haemorrhagic infarction in preterm infants: the role of thrombophilia.* Dev Med Child Neurol, 2012. **54**(2): p. 140-147.

[31] Göpel, W., et al., *Low prevalence of large intraventricular haemorrhage in very low birthweight infants carrying the factor V Leiden or prothrombin G20210A mutation.* Acta Paediatr, 2001. **90**(9): p. 1021-1024.

[32] Debus, O., et al., *Factor V Leiden and genetic defects of thrombophilia in childhood porencephaly.* Arch Dis Child Fetal Neonatal Ed, 1998. **78**(2): p. F121-F124.

[33] Adcock, K., et al., *The TNF-alpha -308, MCP-1 -2518 and TGF-beta1 +915 polymorphisms are not associated with the development of chronic lung disease in very low birth weight infants.* Genes Immun, 2003. **4**(6): p. 420-6.

[34] Bilguvar, K., et al., *COL4A1 mutation in preterm intraventricular hemorrhage.* The Journal of pediatrics, 2009. **155**(5): p. 743-745.

[35] Wang, T., et al., *NT-4 attenuates neuroinflammation via TrkB/PI3K/FoxO1 pathway after germinal matrix hemorrhage in neonatal rats.* J Neuroinflammation, 2020. **17**(1): p. 158.

[36] Vinukonda, G., et al., *Effect of prenatal glucocorticoids on cerebral vasculature of the developing brain.* Stroke, 2010. **41**(8): p. 1766-1773.

[37] Roberts, D. and S. Dalziel, *Antenatal corticosteroids for accelerating fetal lung maturation for women at risk of preterm*

birth. Cochrane Database Syst Rev, 2006(3): p. Cd004454.

[38] Shankaran, S., et al., *Relationship between antenatal steroid administration and grades III and IV intracranial hemorrhage in low birth weight infants. The NICHD Neonatal Research Network*. Am J Obstet Gynecol, 1995. **173**(1): p. 305-312.

[39] Ment, L.R., et al., *Low-dose indomethacin and prevention of intraventricular hemorrhage: a multicenter randomized trial*. Pediatrics, 1994. **93**(4): p. 543-550.

[40] Fowlie, P.W. and P.G. Davis, *Prophylactic indomethacin for preterm infants: a systematic review and meta-analysis*. Arch Dis Child Fetal Neonatal Ed, 2003. **88**(6): p. F464-6.

[41] Schmidt, B., et al., *Long-term effects of indomethacin prophylaxis in extremely-low-birth-weight infants*. N Engl J Med, 2001. **344**(26): p. 1966-1972.

[42] Fowlie, P.W. and P.G. Davis, *Prophylactic intravenous indomethacin for preventing mortality and morbidity in preterm infants*. Cochrane Database Syst Rev, 2002(3): p. Cd000174.

[43] Fowlie, P.W., P.G. Davis, and W. McGuire, *Prophylactic intravenous indomethacin for preventing mortality and morbidity in preterm infants*. Cochrane Database Syst Rev, 2010. **2010**(7): p. Cd000174.

[44] Mohamed, M.A. and H. Aly, *Transport of premature infants is associated with increased risk for intraventricular haemorrhage*. Arch Dis Child Fetal Neonatal Ed, 2010. **95**(6): p. F403-F407.

[45] Crowther, C.A., D.D. Crosby, and D.J. Henderson-Smart, *Vitamin K prior to preterm birth for preventing neonatal periventricular haemorrhage*. Cochrane Database Syst Rev, 2010. **2010**(1): p. Cd000229.

[46] Thorp, J.A., et al., *Antepartum vitamin K and phenobarbital for preventing intraventricular hemorrhage in the premature newborn: a randomized, double-blind, placebo-controlled trial*. Obstet Gynecol, 1994. **83**(1): p. 70-76.

[47] Kaempf, J.W., et al., *Antenatal phenobarbital for the prevention of periventricular and intraventricular hemorrhage: a double-blind, randomized, placebo-controlled, multihospital trial*. J Pediatr, 1990. **117**(6): p. 933-938.

[48] Beverley, D.W., et al., *Prevention of intraventricular haemorrhage by fresh frozen plasma*. Arch Dis Child, 1985. **60**(8): p. 710-713.

[49] Basu, S.K., et al., *Immediate clinical outcomes in preterm neonates receiving antenatal magnesium for neuroprotection*. J Perinat Med, 2011. **40**(2): p. 185-189.

[50] Santhosh, D., et al., *Harnessing region-specific neurovascular signaling to promote germinal matrix vessel maturation and hemorrhage prevention*. Dis Model Mech, 2019. **12**(11).

[51] Romantsik, O., M.G. Calevo, and M. Bruschettini, *Head midline position for preventing the occurrence or extension of germinal matrix-intraventricular hemorrhage in preterm infants*. Cochrane Database Syst Rev, 2017. **7**(7): p. Cd012362.

[52] de Bijl-Marcus, K.A., et al., *The Effect of Head Positioning and Head Tilting on the Incidence of Intraventricular Hemorrhage in Very Preterm Infants: A Systematic Review*. Neonatology, 2017. **111**(3): p. 267-279.

[53] Flores, J.J., et al., *A comprehensive review of therapeutic targets that induce microglia/macrophage-mediated hematoma resolution after germinal*

matrix hemorrhage. J Neurosci Res, 2020. **98**(1): p. 121-128.

[54] Pellicer, A., et al., *Noninvasive continuous monitoring of the effects of head position on brain hemodynamics in ventilated infants.* Pediatrics, 2002. **109**(3): p. 434-440.

[55] Schrod, L. and J. Walter, *Effect of head-up body tilt position on autonomic function and cerebral oxygenation in preterm infants.* Biol Neonate, 2002. **81**(4): p. 255-259.

[56] Schulz, G., et al., *Slow blood sampling from an umbilical artery catheter prevents a decrease in cerebral oxygenation in the preterm newborn.* Pediatrics, 2003. **111**(1): p. e73-e76.

[57] Romantsik, O., M.G. Calevo, and M. Bruschettini, *Head midline position for preventing the occurrence or extension of germinal matrix-intraventricular haemorrhage in preterm infants.* Cochrane Database Syst Rev, 2020. 7(7): p. Cd012362.

[58] Vassilyadi, M., et al., *Functional outcomes among premature infants with intraventricular hemorrhage.* Pediatr Neurosurg, 2009. **45**(4): p. 247-255.

[59] Hennrick, K. and D. Yang, *Hematoidin.* Blood, 2014. **124**(13): p. 2158.

[60] Gulati, R. and M.P. Menon, *Indicators of true intracerebral hemorrhage: hematoidin, siderophage, and erythrophage.* Blood, 2015. **125**(23): p. 3664.

[61] Wysozan, T.R. and R. Gulati, *Revisiting germinal matrix and ventricular lining cells in cerebrospinal fluid: Potential mimickers of intracranial malignancy.* Diagn Cytopathol, 2021. **49**(3): p. 449-451.

[62] Fernandes, S.P. and L. Penchansky, *Tumorlike clusters of immature cells in cerebrospinal fluid of infants.* Pediatr Pathol Lab Med, 1996. **16**(5): p. 721-729.

[63] Fischer, J.R., et al., *Blast-like cells in cerebrospinal fluid of neonates. Possible germinal matrix origin.* Am J Clin Pathol, 1989. **91**(3): p. 255-258.

[64] Kumar, N., et al., *Superficial siderosis.* Neurology, 2006. **66**(8): p. 1144-1152.

[65] Albayram, M.S., et al., *Frequency, Extent, and Correlates of Superficial Siderosis and Ependymal Siderosis in Premature Infants with Germinal Matrix Hemorrhage: An SWI Study.* AJNR Am J Neuroradiol, 2020. **41**(2): p. 331-337.

[66] Strahle, J.M., et al., *Role of hemoglobin and iron in hydrocephalus after neonatal intraventricular hemorrhage.* Neurosurgery, 2014. **75**(6): p. 696-705; discussion 706.

[67] Klebe, D., et al., *Acute and delayed deferoxamine treatment attenuates long-term sequelae after germinal matrix hemorrhage in neonatal rats.* Stroke, 2014. **45**(8): p. 2475-2479.

[68] Li, Q., et al., *Targeting Germinal Matrix Hemorrhage-Induced Overexpression of Sodium-Coupled Bicarbonate Exchanger Reduces Posthemorrhagic Hydrocephalus Formation in Neonatal Rats.* J Am Heart Assoc, 2018. **7**(3).

[69] Zhang, Y., et al., *Bliverdin reductase-A improves neurological function in a germinal matrix hemorrhage rat model.* Neurobiol Dis, 2018. **110**: p. 122-132.

[70] Qing, W.G., et al., *Brain edema after intracerebral hemorrhage in rats: the role of iron overload and aquaporin 4.* J Neurosurg, 2009. **110**(3): p. 462-468.

[71] Tortora, D., et al., *Quantitative susceptibility map analysis in preterm neonates with germinal*

matrix-intraventricular hemorrhage. J Magn Reson Imaging, 2018. **48**(5): p. 1199-1207.

[72] Klebe, D., et al., *Modulating the Immune Response Towards a Neuroregenerative Peri-injury Milieu After Cerebral Hemorrhage*. J Neuroimmune Pharmacol, 2015. **10**(4): p. 576-586.

[73] Chen, S., et al., *An update on inflammation in the acute phase of intracerebral hemorrhage*. Transl Stroke Res, 2015. **6**(1): p. 4-8.

[74] Yang, Y., et al., *Attenuation of acute stroke injury in rat brain by minocycline promotes blood-brain barrier remodeling and alternative microglia/macrophage activation during recovery*. J Neuroinflammation, 2015. **12**: p. 26.

[75] Tang, J., et al., *Minocycline Attenuates Neonatal Germinal-Matrix-Hemorrhage-Induced Neuroinflammation and Brain Edema by Activating Cannabinoid Receptor 2*. Mol Neurobiol, 2016. **53**(3): p. 1935-1948.

[76] Feng, Z., et al., *Anti-inflammation conferred by stimulation of CD200R1 via Dok1 pathway in rat microglia after germinal matrix hemorrhage*. J Cereb Blood Flow Metab, 2019. **39**(1): p. 97-107.

[77] Karimy, J.K., et al., *Inflammation-dependent cerebrospinal fluid hypersecretion by the choroid plexus epithelium in posthemorrhagic hydrocephalus*. Nat Med, 2017. **23**(8): p. 997-1003.

[78] Flores, J.J., et al., *PPARγ-induced upregulation of CD36 enhances hematoma resolution and attenuates long-term neurological deficits after germinal matrix hemorrhage in neonatal rats*. Neurobiol Dis, 2016. **87**: p. 124-133.

[79] Strahle, J., et al., *Mechanisms of hydrocephalus after neonatal and adult intraventricular hemorrhage*. Transl Stroke Res, 2012. **3**(Suppl 1): p. 25-38.

[80] Klebe, D., et al., *Posthemorrhagic hydrocephalus development after germinal matrix hemorrhage: Established mechanisms and proposed pathways*. J Neurosci Res, 2020. **98**(1): p. 105-120.

[81] Manaenko, A., et al., *Inhibition of transforming growth factor-β attenuates brain injury and neurological deficits in a rat model of germinal matrix hemorrhage*. Stroke, 2014. **45**(3): p. 828-834.

[82] Ahn, S.Y., et al., *Mesenchymal stem cells prevent hydrocephalus after severe intraventricular hemorrhage*. Stroke, 2013. **44**(2): p. 497-504.

[83] Ding, Y., et al., *Astrogliosis inhibition attenuates hydrocephalus by increasing cerebrospinal fluid reabsorption through the glymphatic system after germinal matrix hemorrhage*. Exp Neurol, 2019. **320**: p. 113003.

[84] Liu, S.P., et al., *Secukinumab attenuates reactive astrogliosis via IL-17RA/(C/EBPβ)/SIRT1 pathway in a rat model of germinal matrix hemorrhage*. CNS Neurosci Ther, 2019. **25**(10): p. 1151-1161.

[85] Mukerji, A., V. Shah, and P.S. Shah, *Periventricular/Intraventricular Hemorrhage and Neurodevelopmental Outcomes: A Meta-analysis*. Pediatrics, 2015. **136**(6): p. 1132-1143.

[86] Reubsaet, P., et al., *The Impact of Low-Grade Germinal Matrix-Intraventricular Hemorrhage on Neurodevelopmental Outcome of Very Preterm Infants*. Neonatology, 2017. **112**(3): p. 203-210.

[87] Volpe, J.J., *Impaired Neurodevelopmental Outcome After Mild Germinal Matrix-Intraventricular Hemorrhage*. Pediatrics, 2015. **136**(6): p. 1185-1187.

[88] Tortora, D., et al., *The effects of mild germinal matrix-intraventricular haemorrhage on the developmental white matter microstructure of preterm neonates: a DTI study*. Eur Radiol, 2018. **28**(3): p. 1157-1166.

[89] Tortora, D., et al., *Regional impairment of cortical and deep gray matter perfusion in preterm neonates with low-grade germinal matrix-intraventricular hemorrhage: an ASL study*. Neuroradiology, 2020. **62**(12): p. 1689-1699.

[90] Back, S.A., et al., *Late oligodendrocyte progenitors coincide with the developmental window of vulnerability for human perinatal white matter injury*. J Neurosci, 2001. **21**(4): p. 1302-1312.

[91] Xu, G., et al., *Late development of the GABAergic system in the human cerebral cortex and white matter*. J Neuropathol Exp Neurol, 2011. **70**(10): p. 841-858.